SPACES OF HOPE

D1203006

California Studies in Critical Human Geography

Editorial Board:

Professor Michael Watts, University of California · Berkeley
Professor Allan Pred, University of California · Berkeley
Professor Richard Walker, University of California · Berkeley
Professor Gillian Hart, University of California · Berkeley
Professor AnnaLee Saxenian, University of California · Berkeley
Professor Mary Beth Pudup, University of California · Santa Cruz

Spaces of Hope

David Harvey

University of California Press
Berkeley   Los Angeles

for Delfina and Her Generation

Hope is Memory that Desires

(Balzac)

© David Harvey, 2000

University of California Press
Berkeley and Los Angeles, California

Published by arrangement with
Edinburgh University Press

Typeset in Ehrhardt by
Bibliocraft Ltd, Dundee.
Printed and bound in Great Britain by
The Cromwell Press, Trowbridge, Wilts

ISBN 0-520-22577-5 (cloth)
ISBN 0-520-22578-3 (paper)

9   8   7   6   5   4   3   2   1

Contents

Acknowledgements

I have drawn extensively on materials published (or about to be published) in a variety of places. The original sources include, 'The geography of class power,' *The Socialist Register*, 1998, 49–74); 'Globalization in question,' *Rethinking Marxism*, 8, 1996, 1–17; 'The body as an accumulation strategy,' *Society and Space*, 16, 1998, 401–21; 'Considerations on the environment of justice,' forthcoming in Low, N. (ed.), *Global Ethics*, Routledge, London; 'The spaces of utopia,' forthcoming in Bowers, L., Goldberg, D., and Mushenyo, M. (eds), *Between Law and Society*, Minnesota Press; 'Marxism, metaphors, and ecological politics,' *Monthly Review*, April 1998, 17–31; 'Frontiers of insurgent planning,' to be published in the journal *Plurimondi*. I want to thank the editors and the referees of some of these pieces for their helpful comments. I have also used materials from 'The work of postmodernity: The body in global space,' a paper first delivered to The Postmodernity Project in the University of Virginia. I have, furthermore, been privileged to receive considerable feedback in a variety of institutional settings such as a powerful workshop on urban alternatives in Perugia (organized by Leonie Sandercock and Dino Borri), a week long seminar at the Tapies Foundation in Barcelona (thanks particularly to Noemi Cohen), a visit with the Justice Studies Center in Arizona State University, The Humanities Center in the University of Oregon, and many others too numerous to mention. Individual, group, and audience responses to seminars, talks, lectures, and discussions have been important in shaping my thinking and I want to thank all those who have contributed to that process.

There are, however, some individuals who have helped in more specific ways. Cindi Katz and Neil Smith have been marvelously supportive. A variety of people have contributed (often, I think, without knowing it) to my thinking on specific topics. These include Jonathan Lange, Bertell Ollman, Peter Gould, Neil Hertz, Bill Leslie, Mark Blyth, Emily Martin, Katherine Verdery, Reds Wolman, Erik Swyngedouw, Andy Merrifield, Melissa Wright, Haripriya Rangan, Jean-Francois Chevrier, Brian

Holmes, Masao Miyoshi, and a talented group of students at Hopkins (both graduate and undergraduate), particularly those who worked so hard to keep the living-wage campaign in the city and the University so directly alive. I particularly want to thank Mark Damien who helped identify and procure some of the illustrative materials. And I owe (as always) a great debt to John Davey, now of Edinburgh University Press, for his editorial wisdom and encouragement. I wish him well in his new venture. Finally, Haydee and Delfina always help define special perspectives on life and loving that make the usual difficulties of authorship seem much more bearable.

Figure acknowledgements

Plate 3.1   is a visualization by Penny Masuoka, UMBC, NASA God-dard Space Flight Center and William Acevedo, USGS, NASA Ames Research Center.

Plate 8.17   *View of an Ideal City* is reproduced with permission from the Walters Art Gallery, Baltimore.

Plate 8.18   is reproduced from *Utopia* by Thomas More, 1518, Shelf-mark, Mar. 89, is reproduced by courtesy of the Bodleian Library, University of Oxford.

Plate 8.23   Fondation Le Corbusier, Paris, © DACS 1988 and Metro-politan Life Insurance Company Archives, New York.

Plate 8.24   is reproduced with permission © 1999 The Frank Lloyd Wright Foundation, Scottsdale, AZ.

Plate 8.25   Poundbury, Dorset, is reproduced by courtesy of the Duchy of Cornwall. Photo by Mark Fiennes.

List of plates

INTRODUCTION

CHAPTER 1

The difference a generation makes

1 Marx redux

Every year since 1971 (with the exception of one) I have run either a reading group or a course on Marx's *Capital* (Volume 1). While this may reasonably be taken as the mark of a peculiarly stodgy academic mind, it has allowed me to accumulate a rare time-series of reactions to this particular text.

In the early 1970s there was great political enthusiasm for it on the part of at least a radical minority. Participation was understood as a political act. Indeed, the course was set up (in parallel with many others of its sort across American campuses at the time) to try to find a theoretical basis, a way of understanding all of the chaos and political disruption evident in the world (the civil rights movement of the 1960s and the urban uprisings that followed the assassination of Martin Luther King in the United States, the growing opposition to the imperialist war in Vietnam, the massive student movements of 1968 that shook the world from Paris to Mexico City, from Berkeley and Berlin to Bangkok, the Czech 'Spring' and its subsequent repression by the Soviets, the 'Seven Days War' in the Middle East, the dramatic events that occurred at the Democratic National Convention in Chicago, just to name a few of the signal events that made it seem as if the world as we knew it was falling apart).

In the midst of all this turmoil there was a crying need for some sort of political and intellectual guidance. Given the way in which Marx's works had effectively been proscribed through the long history of McCarthyite repression in the United States, it seemed only right and proper to turn to Marx. He must have had something important to say, we reasoned, otherwise his works would not have been suppressed for so long. This presumption was given credibility by the icy reception to our efforts on many a campus. I disguised the name of the course, often ran it of an evening and gave 'independent study' credit for those who did not want any mention of it on their transcript (I later learned from someone high up in the administration that since the course was taught in the geography

3

program and was called 'Reading Capital' it took them nearly a decade to figure out it was Marx's *Capital* that was being taught!).

Capital was not an easy text to decipher, at least for the uninitiated (and there were many of us in that condition and only a few old hands could help us on our way – most of them of European extraction where communist parties had long remained active). But for those of us in universities the intellectual difficulty was, at least, a normal challenge.

In these early years many young faculty members participated as did many graduate students. Some of them have gone on to be famous (and though some have changed their stripes most will generously acknowledge the formative nature of the whole experience). They came from all manner of disciplines (Philosophy, Math Sciences, Political Theory, History of Science, English, Geography, History, Sociology, Economics ...). In retrospect I realize what an incredible privilege it was to work through this text with people armed with so many different intellectual skills and political perspectives. This was how I learned my Marx, through a process of mutual self-education that obeyed little or no particular disciplinary logic let alone party political line. I soon found myself teaching the text well beyond the confines of the university, in the community (with activists, teachers, unionists). I even got to teach some of it (not very successfully) in the Maryland penitentiary.

Teaching undergraduates was somewhat more fraught. The dominant tone of undergraduate radicalism in those days was anti-intellectual. For them, the academy seemed the center of ideological repressions; book learning of any sort was inherently suspect as a tool of indoctrination and domination. Many undergraduate student activists (and these were, of course, the only ones who would ever think of taking the course) thought it rather unradical to demand that they read let alone understand and write about such a long and tortuous book. Not many of them lasted the course. They paid no mind to Marx's injunction that 'there is no royal road to science' nor did they listen to the warning that many readers 'always anxious to come to a conclusion, eager to know the connexion between general principles and the immediate questions that have aroused their passions, may be disheartened because they will be unable to move on at once.' No amount of 'forewarning and forearming those readers who zealously seek the truth' (Marx, 1976 edition, 104) seemed to work with this audience. They were carried forward largely on a cresting wave of intuitions and bruised emotions (not, I hasten to add, necessarily a bad thing).

The situation is radically different now. I teach *Capital* purely as a respectable regular course. I rarely if ever see any faculty members and the graduate student audience has largely disappeared (except for those few who plan to work with me and who take the course as some kind of 'rite of passage' before they go on to more important things). Most of the graduate survey courses in other departments now allot Marx a week or two, sandwiched in between, say, Darwin and Weber. Marx gets attention. But in academia, this is devoted either to putting him in his place as, say, a 'minor post-Ricardian' or passing him by as an out-moded 'structuralist' or 'modernist.' Marx is, in short, largely written off as the weaver of an impossibly huge masternarrative of history and an advocate of some totally impossible historical transformation that has in any case been proven by events to be just as fallacious politically and practically as it always was theoretically.

Even before the collapse of the Berlin Wall, in the early 1980s, Marx was definitely moving out of academic and political fashion. In the halcyon years of identity politics and the famous 'cultural turn' the Marxian tradition assumed an important negative role. It was ritualistically held up (incorrectly) as a dominant ideology that had to be fought against. Marx and 'traditional' Marxism were systematically criticized and denigrated as insufficiently concerned with more important questions of gender, race, sexuality, human desires, religion, ethnicity, colonial dominations, environment, or whatever. Cultural powers and movements were just as important if not more so than those of class and what was class anyways if not one out of many different and cross-cutting cultural configurations. All of that might have been fair enough (there were plenty of grounds for such criticisms) if it had not also been concluded that Marxism as a mode of thought was inherently antagonistic towards any such alternative formulations and therefore a totally lost cause. In particular, cultural analysis supplanted political economy (the former, in any case, being much more fun than being absorbed in the dour world and crushing realities of capitalist exploitation).

And then came the collapse of the Wall, the last nail in the coffin of any sort of Marxist credibility even if many of a Marxian persuasion had long distanced themselves (some as long ago as the Hungarian uprising of 1956 and still more with the crushing of the Czech Spring in 1968) from actually existing socialism of the Soviet-Chinese sort. To pretend there was anything interesting about Marx after 1989 was to sound more and more like an all-but extinct dinosaur whimpering its own last rites. Free-market capitalism rode triumphantly across the globe, slaying all such old dinosaurs in its path. 'Marx talk' was increasingly confined to what might best be described as an increasingly geriatric 'New Left' (I myself passed none too gently into that night known as 'senior citizen'). By the early 1990s the intellectual heft of Marxian theory seemed to be terminally in decline.

But some undergraduates still continue to take the *Capital* course. For most of them this is no longer a political act. The fear of communism has largely dissipated. The course has a good reputation. A few students are curious to see what all the fuss with Marxism was about. And a few still have some radical instincts left to which they feel Marx might add an extra insight or two. So, depending on their timetable and their requirements, some undergraduates end up in Marx's *Capital* rather than in Aristotle's *Ethics* or Plato's *Republic*.

This contrast I have drawn between then and now in terms of political and intellectual interest and response to Marx is hardly surprising. Most will recognize the broad outlines of what I have described even if the specific lense I am using exaggerates and distorts here and there.

But there is another tale to be told that makes matters rather more confusing. In the early 1970s it was hard to find the direct relevance of Volume 1 of *Capital* to the political issues that dominated the day. We needed Lenin to get us from Marx to an understanding of the imperialist war that so unnerved us in Vietnam. We needed a theory of civil society (Gramsci at least) to get us from Marx to civil rights, and a theory of the state (such as Miliband or Poulantzas) to get us to a critique of state repressions and welfare state expenditures manipulated to requirements of capital accumulation. We needed the Frankfurt School to understand questions of legitimacy, technological rationality, the state and bureaucracy, and the environment.

But then consider the historical-geographical conditions. In much of the advanced capitalist world, the trade union movement (often far too reformist for our radical tastes) was still strong, unemployment was broadly contained, everywhere (except in the United States) nationalization and public ownership was still on the agenda and the welfare state had been built up to a point where it seemed unassailable if flawed. Elsewhere in the world movements were afoot that seemed to threaten the existence of capitalism. Mao was a pre-eminent revolutionary leader in China while many other charismatic revolutionaries from Che Guevara and Castro in the Latin American context to Cabral and Nyerere in Africa actively held out the possibility of a socialist or communist alternative.

Revolution seemed imminent and we have subsequently learned that it was actively feared among many of the rulers of the time (even going beyond what might be expected from the evident paranoia of someone like Richard Nixon). How that revolution might occur and the kind of society to which it might lead were not topics even remotely touched upon in Marx's *Capital* (though there were plenty of other texts of Marx and the Marxists to which we could turn for enlightenment).

In short, we needed a whole host of mediations to get from Marx's *Capital* to the political issues that concerned us. And it frequently entailed an act of faith in the whole history of the Marxist movement (or in some charismatic figure like Mao or Castro) to believe in the inner connection between Marx's *Capital* and all that we were interested in. This is not to say there was nothing in the text to fascinate and delight – the extraordinary insights that came from consideration of the commodity fetish, the wonderful sense of how class struggle had altered the world from the pristine forms of capital accumulation that Marx described. And once one got used to it, the text provided its own peculiar and beguiling pleasures. But the plain fact was that *Capital* did not have that much direct relevance to daily life. It described capitalism in its raw, unmodified, and most barbaric nineteenth-century state.

The situation today is radically different. The text teems with ideas as to how to explain our current state. There is the fetish of the market that caught out that lover of children Kathy Lee Gifford when she was told that the line of clothing she was selling through Wal-Mart was made either by thirteen-year-olds in Honduras paid a mere pittance or by sweated women workers in New York who had not been paid for weeks. There is also the whole savage history of downsizing (prominently reported on in the *New York Times*), the scandals over child labor in Pakistan in the manufacture of carpets and soccer balls (a scandal that was forced upon FIFA's attention), and Michael Jordan's $30 million retainer for Nike, set against press accounts of the appalling conditions of Nike workers in Indonesia and Vietnam. The press is full of complaints as to how technological change is destroying employment opportunities, weakening the institutions of organized labor and increasing rather than lightening the intensity and hours of labor (all central themes of Marx's chapter on 'Machinery and Modern Industry'). And then there is the whole question of how an 'industrial reserve army' of labor has been produced, sustained and manipulated in the interests of capital accumulation these last decades, including the public admission by Alan Budd, an erstwhile advisor to Margaret Thatcher, that the fight against inflation in the early 1980s was a cover for raising unemployment and reducing the strength of the working class. 'What was engineered,' he said, 'in Marxist terms – was a crisis in capitalism which re-created a reserve army of labour, and has allowed the capitalists to make high profits ever since' (Brooks, 1992).

All of this now makes it all too easy to connect Marx's text to daily life. Students who stray into the course soon feel the heat of what amounts to a devastating critique of a world of free-market neoliberalism run riot. For their final paper I give them bundles of cuttings from the *New York Times* (a respectable source, after all) and suggest they use them to answer an imaginary letter from a parent/relative/friend from home that says:

> I hear you are taking a course on Marx's Das Kapital. I have never read it myself though I hear it is both interesting and difficult. But thank heavens we have put that nineteenth century nonsense behind us now. Life was hard and terrible in those days, but we have come to our collective senses and made a world that Marx would surely never recognize . . .

They write illuminating and often devastatingly critical letters in reply. Though they dare not send them, few finish the course without having their views disrupted by the sheer power of a text that connects so trenchantly with conditions around us.

Herein, then, lies a paradox. This text of Marx's was much sought after and studied in radical circles at a time when it had little direct relationship to daily life. But now, when the text is so pertinent, scarcely anyone cares to consider it. Why?

2 The difference a generation makes

I found myself asking the same question, though from a completely different angle, when I happened to view two films back to back (courtesy of capitalistic video technology). I saw them in reverse chronological order and I suspect that the impact they made on me was all the stronger for it. The two films were *Hate* (La Haine) which came out in 1995, and Jean-Luc-Godard's classic piece from 1966 called *One or Two Things I Know about Her* (Un ou deux choses que je sais d'elle).

Hate records a day in the life of three young men. Two are children of Maghrebian and African immigrants respectively and the third is of Jewish extraction. Their bond arises from the conditions of life of contemporary youth raised in the suburban projects (the public housing projects built for workers largely during the 1960s). Together, they face a world of unemployment, police repression, arbitrary state power, social breakdown, and loss of any sense of belonging or citizenship. The urban and (in this instance) suburban unrest, coupled with violent confrontations and street fights, lootings and burnings that periodically occurred in many French cities in the 1990s form the background to the story. This backdrop could be extended to include the violence that followed the Rodney King verdict in Los Angeles in 1992, the sudden outbreaks of youth violence in Manchester, Liverpool, and even 'nice' cities, like Oxford, in the late 1980s, as well as eruptions in several European metropolitan areas.

The film is full of raw anger, pain, and violent despair on the part of the three main protagonists. They cannot even have kind words for each other, let alone for anyone else. There is scarcely a moment of tenderness let alone reflective or thoughtful examination in the whole film ('thinking too much' is a term of derision). Driven by their anger and raw emotions, these individuals are vulnerable to the core. They seem stripped of all defenses, yet they also desperately seek attention, identity, and recognition by engaging in the only kind of behavior that draws the attention of those possessed of power – sometimes sullen but always unpredictable and disruptive aggression. The only form of empowerment available to the protaganists lies in the gun (a service weapon lost by a police officer and found by one of them). The only relevant existential question that hangs over the film is how and when to use that gun.

The film itself mirrors this transgressive behavior in its technique and form. No attention is paid to subtleties. The film is as raw, crude, and unreflective as its subject matter. It uses the same techniques to attract attention as its subjects do. This is, the film seems to say, the contemporary metropolis at work. A place of both artistic and lived impoverishment if not humane impossibility.

Godard's film opens with the sounds and scenes of construction. Such scenes form a frequent interruptor of processes of quiet reflection on the part of the masculine narrator (the director) and the actresses who talk about their lives and give of their thoughts as they pass their day, prostituting themselves both to men as well as to the icons of contemporary culture (the car, the boutique, the highways, the suburban public housing projects, the simplistic bourgeois version of family life – the film in fact ends with a shot of an urban landscape made up of a bundle of consumer products). The city is a space in the process of formation. It is already marked by the traces of deindustrialization and the rise of the political economy of the sign as opposed to a political economy of direct material reproduction. The question mark that hovers over it is what will become of the inhabitants as this new urban world is created. By the time we view *Hate*, the answer to that question is omnipresent, but in 1966 Godard builds a subtle sense of dread, a sense of getting lost and of fragmentation. Yet there is a strong sub-text that hints of alternatives. Can we recover the ABC of existence, the narrator asks?

The pieces seem to hold together as a solid reflection on what the city can and might mean (for the figure of 'the city' looms large throughout the film). But the individuals seem caught, helplessly passive, imprisoned and fragmented within the web of urban life being constructed by agents of power that seem far away – the Gaullist state with its all-powerful planning agencies, monopoly capital in cahoots with that state to rebuild a world and the city in its own image, and beyond that the globalizing reach of the United States waging a Cold War against Moscow and Beijing and a hot war in Vietnam, dominating the media, and placing signs of its power everywhere (TWA, PAN AM . . .). But even that contains a possibility, however problematic. 'Suddenly I had the impression that I was the world and the world was me,' says the alienated Juliette as she surveys the bland facades of the new residential tower blocks rising in the Paris suburbs.

The movie operates at a key point in the emergence of what later came to be known as a postmodern sensibility. It poses all the questions we are now all too familiar with. The limits of language (Wittgenstein's 'the limits of my language mean the limits of my world' is directly quoted), the impossibility of 'true' communication, the feeling that something is missing ('but I know not what'), the inability to represent events in all their fullness, the sense that 'life is like a comic strip' and the perverse ways in which signs, representations, and language confuse rather than clarify an always elusive reality. The physical clarity of images and representations, in the film as in the commentary itself, contrasts with the murkiness of a future in which the only hope seems to lie in the capacity of the human spirit somehow to 'take possession' of minor things and 'to catch a fleeting reason to be alive.' So what are dreams made of in such a world? Says Juliette in answer to that question (posed by her young son) 'I used to dream I was being sucked into a huge hole but now I feel I am being scattered into a thousand pieces and when I wake I worry that a piece is missing.' The fearful paranoia often associated with modernity here gives way to the sensibility of schizophrenia so often associated with postmodernity. Existential and phenomenological sensibilities (long present on the French Left) filtered through Marxism point (as in the parallel case of Althusserian philosophy, at its apogee when Godard's film was produced) towards a fragmented and postmodernist way of thinking.

Yet the film, as Chevrier (1997) points out, is a thing of enormous and compelling beauty. Like the calm and gentle beauty of its actress, the film uses aesthetic powers and intellectuality as a defense against pain. In this it counters the passivity of its characters with a subterranean utopian activism. It conjures a sense of future possibilities out of nothingness simply by virtue of the wide-ranging questions it poses. 'If things come into focus again,' says the narrator, 'this can only be through the rebirth of *conscience*' (the latter word having multiple meanings in French that stretch from external to inner understandings). It is the power of the human spirit rather than of the gun that holds the key to the future.

Godard, the avant-garde leftist film director, articulates the problematics of French Maoism and Althusserianism and in so doing pioneers the transition to postmodernism through an artistic and intellectual *tour de force* that many now acknowledge helped open the flood gates on the left to a new mode of radical thought. In the short run this produced the radical movement of 1968 but in the long run it led to the demise of functionalist, dogmatic, and foundationalist forms of Marxism. The latter proved inappropriate to the complicated world of monopoly state capitalism coupled with the post-war welfare state and a rising consumer culture in which the political economy of the spectacle and the sign were to take on new and enhanced roles.

But those times have changed. The city of the future that hovers as a question mark in Godard's film is fully formed in *Hate*. Utopian longing has given way to unemployment, discrimination, despair, and alienation. Repressions and anger are now everywhere in evidence. There is no intellectual or aesthetic defense against them. Signs don't even matter in any fundamental sense any more. The city incarcerates the underprivileged and further marginalizes them in relation to the broader society. But is not this exactly the kind of world for which a rather traditional, even crude and (dare I say it) vulgar and functionalist Marxism might be all too politically appropriate? What would happen if the 'penurious rabble' that Hegel deemed so threatening to the stability of civil society became a 'dangerous class' for itself? There is, of course, no hint of such a turn in the film (though the fascist skinheads bring echoes of a resurrected older power to be struggled against). But the parallel to my teaching experience with Marx is striking. Godard struggled to free himself from the chains of a dogmatic Marxism at a certain historical moment while keeping faith with some kind of Marxist/Maoist future. *Hate* records the absence of any such politics in a time and a place where some version of it should surely be appropriate.

3 The work of postmodernity

The paradoxes I have described relate to a massive discursive shift that has occurred over the past three decades. There are all kinds of aspects to this shift and it is easy to get lost in a mass of intricacies and complexities. But what is now striking is the dominance of an almost fairy-tale like belief, held on all sides alike, that once upon a time there was structuralism, modernism, industrialism, Marxism or what have you and now there is post-structuralism, postmodernism, postindustrialism, post-Marxism, post-colonialism, and so forth. Like all such tales, this one is rarely spoken of in such a crude or simplistic way. To do so would be particularly embarrassing to those who deny in principle the significance of broad-based 'metanarratives.' Yet the prevalence of 'the post' (and the associated inability to say what it is that we might be 'pre') is a dominant characteristic of contemporary debate. It has also become a serious game in academia to hunt the covert modernists (if you are a dedicated postmodernist) or to hunt the decadent postmodernists (if you happen to be in favor of some sort of modernist revival).

One of the consequences of this prevalent fairy tale (and I call it that to capture its beguiling power) is that it is impossible to discuss Marx or Marxism outside of these dominant terms of debate. For example, one quite common reaction to my recent work, particularly *Justice, Nature and the Geography of Difference,* is to express surprise and disbelief at how I seem to merge modernist and postmodernist, structuralist and post-structuralist arguments (see, e.g., Eagleton, 1997). But Marx had not read Saussure or Lévi Strauss and while there are some powerful structuralist readings of Marx (principally by Althusser) the evidence that Marx was a structuralist or even a modernist *avant la lettre*, as these terms came to be understood in the 1970s, is neither overwhelming nor conclusive. Analyses based on Marx's work collide with the beguiling power of this fairy-tale reading of our recent discursive history. Put bluntly, we do not read Marx these days (no matter whether he is relevant or not) because he is someone whose work lies in a category that we are supposed to be 'post'. Or if we do read him, it is solely through the lenses provided by what it is we believe we are 'post.'

Now it is indeed interesting to look at Marx's *ouevre* through such lenses. He was, of course, an avid critic of classical bourgeois political economy and devoted much of his life to 'deconstructing' its dominant principles. He was deeply concerned with language (discourse) and was acutely aware of how discursive shifts (of the sort he examined in depth in *The Eighteenth Brumaire*) carried their own distinctive political freight. He understood in a deep sense the relationship between knowledge and 'situatedness' ('positionality') though it was, of course, the 'standpoint' of the worker that was the focus of his attention. I could go on and on in this vein, but my point here is not to try to prove that much of what passes for innovative in our recent discursive history is already pre-figured in Marx, but to point to the damage that the fairy-tale reading of the differences between the 'then' and the 'now' is doing to our abilities to confront the changes occurring around us. Cutting ourselves off from Marx is to cut off our investigative noses to satisfy the superficial face of contemporary intellectual fashion.

Bearing this in mind, let me now focus on two facets of this discursive shift that have occurred since around 1970: those captured through the terms 'globalization' and 'the body.' Both terms were little if at all in evidence as analytical tools in the early 1970s. Both are now powerfully present; they can even be regarded as conceptual dominants. 'Globaliza-tion,' for example, was entirely unknown before the mid-1970s. Innumer-able conferences now study the idea. There is a vast literature on the subject, coming at it from all angles. It is a frequent topic of commentary in the media. It is now one of the most hegemonic concepts for under-standing the political economy of international capitalism. And its uses extend far beyond the business world to embrace questions of politics, culture, national identity, and the like. So where did this concept come from? Does it describe something essentially new?

'Globalization' seems first to have acquired its prominence as American Express advertised the global reach of its credit card in the mid-1970s. The term then spread like wildfire in the financial and business press, mainly as legitimation for the deregulation of financial markets. It then helped make the diminution in state powers to regulate capital flows seem inevitable and became an extraordinarily powerful political tool in the disempowerment of national and local working-class movements and trade union power (labor discipline and fiscal austerity – often imposed by the International Monetary Fund and the World Bank – became essential to achieving internal stability and international competitiveness). And by the mid-1980s it helped create a heady atmosphere of entrepreneurial optimism around the theme of the liberation of markets from state control. It became a central concept, in short, associated with the brave new world of globalizing neoliberalism. It helped make it seem as if we were entering upon a new era (with a touch of teleological inevitability thrown in) and thereby became part of that package of concepts that distinguished between then and now in terms of political possibilities. The more the left adopted this discourse as a description of the state of the world (even if it was a state to be criticized and rebelled against), the more it circum-scribed its own political possibilities. That so many of us took the concept on board so uncritically in the 1980s and 1990s, allowing it to displace the far more politically charged concepts of imperialism and neocolonialism, should give us pause. It made us weak opponents of the politics of glob-alization particularly as these became more and more central to everything that US foreign policy was trying to achieve. The only politics left was a politics of conserving and in some instances downright conservative resistance.

There is, however, one other angle on much of this that may have equally deep significance. The NASA satellite image entitled 'Earth Rise' depicted the earth as a free-floating globe in space. It quickly assumed the status of an icon of a new kind of consciousness. But the geometrical properties of a globe are different from those of a two-dimensional map. It has no natural boundaries save those given by lands and oceans, cloud covers and vegetation patterns, deserts and well-watered regions. Nor does it have any particular center. It is perhaps no accident that the awareness of the artificiality of all those boundaries and centers that had hitherto dominated thinking about the world became much more acute. It became much easier, with this icon of the globe hanging in the background, to write of a 'borderless world' (as Miyoshi, 1997, has so persuasively done) and to take a radically decentered approach to culture (with the massive cultural traditions of China, India, South America, and Africa suddenly looking as salient and as geographically dominant across segments of the globe as those of the West). Travel around the world, already much easier, suddenly had no natural stopping point and the continuity of spatial relations suddenly becomes both practically and rhetorically a fundamental fact of life. And it may well be that the focus on the body as the center of all things is itself a response to this decentering of everything else, promoted by the image of the globe (rather than the two-dimensional map) as the locus of human activity and thought.

So what of the body? Here the tale, though analogous, is substantially different. The extraordinary efflorescence of interest in 'the body' as a grounding for all sorts of theoretical enquiries over the last two decades has a dual origin. In the first place, the questions raised particularly through what is known as 'second-wave feminism' could not be answered without close attention being paid to the 'nature–nurture' problem and it was inevitable that the status and understanding of 'the body' became central to theoretical debate. Questions of gender, sexuality, the power of symbolic orders, and the significance of psychoanalysis also repositioned the body as both subject and object of discussion and debate. And to the degree that all of this opened up a terrain of enquiry that was well beyond traditional conceptual apparatuses (such as that contained in Marx), so an extensive and original theorizing of the body became essential to progressive and emancipatory politics (this was particularly the case with respect to feminist and queer theory). And there is indeed much that has been both innovative and profoundly progressive within this movement.

The second impulse to return to the body arose out of the movements of post-structuralism in general and deconstruction in particular. The effect of these movements was to generate a loss of confidence in all previously established categories (such as those proposed by Marx) for understanding the world. And it is in this context that the connexion between decentering and the figure of the globe may have done its undermining work. The effect, however, was to provoke a return to the body as the irreducible basis for understanding. Lowe (1995, 14) argues that:

> [T]here still remains one referent apart from all the other destabilized referents, whose presence cannot be denied, and that is the body referent, our very own lived body. This body referent is in fact the referent of all referents, in the sense that ultimately all signifieds, values, or meanings

refer to the delineation and satisfaction of the needs of the body. Precisely because all other referents are now destabilized, the body referent, our own body, has emerged as a problem.

The convergence of these two broad movements has refocussed attention upon the body as the basis for understanding and, in certain circles at least (particularly those animated by writers such as Foucault and Judith Butler), as the privileged site of political resistance and emancipatory politics.

I will shortly take up 'globalization' and 'the body' in greater detail. But I here want merely to comment on the positioning of these two discursive regimes in our contemporary constructions. 'Globalization' is the most macro of all discourses that we have available to us while that of 'the body' is surely the most micro from the standpoint of understanding the work-ings of society (unless, that is, we succumb to the reductionism of seeing society as merely an expression of DNA codings and genetic evolutions). These two discursive regimes – globalization and the body – operate at opposite ends of the spectrum in the scalar we might use to understand social and political life. But little or no systematic attempt has been made to integrate 'body talk' with 'globalization talk.' The only strong connec-tions to have emerged in recent years concern individual and human rights (e.g. the work of Amnesty International), and, more specifically, the right of women to control their own bodies and reproductive strategies as a means to approach global population problems (dominant themes in the Cairo Conference on Population in 1994 and the Beijing Women's Con-ference of 1996). Environmentalists often try to forge similar connections, linking personal health and consumption practices with global problems of toxic waste generation, ozone depletion, global warming, and the like. These instances illustrate the potency and the power of linking two seemingly disparate discursive regimes. But there is a large untilled terrain within which these discursive regimes have been conveniently separated from each other. In this book, therefore, I sketch in a way in which 'globalization' and 'the body' might be more closely integrated with each other and explore the political-intellectual consequences of making such a connexion.

The line of argument I shall use is broadly based in a relational conception of dialectics embodied in the approach that I have come to call 'historical-geographical materialism.' I want, at the outset, to lay out just one fundamental tenet of this approach in order to lay another of the key shibboleths of our time as firmly to rest as I can. And this concerns the tricky question of the relation between 'particularity' and 'universality' in the construction of knowledge.

I deny that we have a choice between particularity or universality in our mode of thinking and argumentation. Within a relational dialectics one is always internalized and implicated in the other. There is a link between, for example, the particularities of concrete labors occurring in particular places and times (the seamstress in Bangladesh who made my shirt), and the measured value of that labor arrived at through processes of exchange, commodification, monetization, and, of course, the circulation and accumulation of capital. One conception of labor is concrete and particular and the other is abstract and 'universal' (in the sense that it is achieved through specific processes of generalization).

Obviously, there could be no abstract labor at all without a million and one concrete labors occurring throughout the world. But what is then interesting is the way in which the qualities of concrete labor respond and internalize the force of abstract labor as achieved through global trade and interaction. Workers engaging in productive concrete labors suddenly find themselves laid off, downsized, rendered technologically obsolete, forced to adapt to new labor processes and conditions of work, simply because of the force of competition (or, put in the terms proposed here, the concrete labor adjusts to abstract conditions at the same time as the qualities of abstract labor depend upon movements and transitions in concrete labor processes in different places and times).

I have used this example to illustrate a general point. The particularity of the body cannot be understood independently of its embeddedness in socio-ecological processes. If, as many now argue, the body is a social construct, then it cannot be understood outside of the forces that swirl around it and construct it. One of those key determinants is the labor process, and globalization describes how that process is being shaped by political-economic and associated cultural forces in distinctive ways. It then follows that the body cannot be understood, theoretically or empirically, outside of an understanding of globalization. But conversely, boiled down to its simplest determinations, globalization is about the socio-spatial relations between billions of individuals. Herein lies the foundational connexion that must be made between two discourses that typically remain segregated, to the detriment of both.

Part of the work of postmodernity as a set of discursive practices over the last two decades has been to fragment and sever connexions. In some instances this proved a wise, important, and useful strategy to try to unpack matters (such as those of sexuality or the relation to nature) that would otherwise have remained hidden. But it is now time to reconnect. This book is an account of what happens when we try to do so.

There is a final point that I need to make. One important root of the so-called 'cultural turn' in recent thinking lies in the work of Raymond

Williams and the study of Gramsci's writings (both particularly important to the cultural studies movement that began in Birmingham with Stuart Hall as one of its most articulate members). One of the several strange and unanticipated results of this movement has been the transformation of Gramsci's remark on 'pessimism of the intellect and optimism of the will' into a virtual law of human nature. I wish in no way to detract from the extraordinary feats of many on the left who have fought a rearguard action against the wave of neoliberalism that swept across the advanced capitalist world after 1980. This showed optimism of the will at its noble best. But a powerful inhibitor to action was the inability to come up with an alternative to the Thatcherite doctrine that 'there is no alternative' (a phrase that will echo as a recurring refrain throughout this book). The inability to find an 'optimism of the intellect' with which to work through alternatives has now become one of the most serious barriers to progressive politics.

Gramsci penned those famous words while sick and close to death in an Italian prison cell under conditions that were appalling. I think we owe it to him to recognize the contingent nature of the comment. We are not in prison cells. Why, then, might we willingly choose a metaphor drawn from incarceration as a guiding light for our own thinking? Did not Gramsci (1978, 213) also bitterly complain, before his incarceration, at the pessimism which produced then the same political passivity, intellectual torpor and scepticism towards the future as it does now in ours? Do we not also owe it to him, out of respect for the kind of fortitude and political passion he exhibited, to transform that phrase in such a way as to seek an optimism of the intellect that, properly coupled with an optimism of the will, might produce a better future? And if I turn towards the end of this book towards the figure of utopia and if I parallel Raymond Williams's title *Resources of Hope* with the title *Spaces of Hope*, then it is because I believe that in this moment in our history we have something of great import to accomplish by exercising an optimism of the intellect in order to open up ways of thinking that have for too long remained foreclosed.

1998 is, it turns out, a fortuitous year to be writing about such matters. It is the thirtieth anniversary (the usual span given to a generation) of that remarkable movement that shook the world from Mexico City to Chicago, Berlin and Paris. More locally (for me), it is thirty years now since much of central Baltimore burned in the wake of the riots that followed the assassination of Martin Luther King (I moved from Bristol to Baltimore the year after that). If only for these reasons this is, therefore, a good moment to take stock of that generational shift that I began by reflecting upon.

But 1998 is also the 150th anniversary of the publication of that most extraordinary of all documents known as *The Communist Manifesto*. And it happens to be the 50th anniversary of the signing of the *Universal*

Declaration of Human Rights at the United Nations. Connecting these events and reflecting on their general meaning appears a worthwhile way to reflect on our contemporary condition. While Marx was deeply suspicious of all talk about rights (sensing it to be a bourgeois trap), what on earth are workers of the world supposed to unite about unless it is some sense of their fundamental rights as human beings? Connecting the sentiments of the *Manifesto* with those expressed in the *Declaration of Human Rights* provides one way to link discourses about globalization with those of the body. The overall effect, I hope, is to redefine in a more subtle way the terms and spaces of political struggle open to us in these extraordinary times.

PART 1

UNEVEN GEOGRAPHICAL DEVELOPMENTS

CHAPTER 2

The geography of the *Manifesto*

What we now call 'globalization' has been around in some form or another for a very long time – at least as far back as 1492 if not before. The phenomenon and its political-economic consequences have likewise long been the subject of commentary, not least by Marx and Engels who, in *The Manifesto of the Communist Party*, published an impassioned as well as thorough analysis of it as long ago as 1848. Workers of the world would have to unite in struggle, they concluded, if they were to subdue the destructive powers of capital on the world stage and construct an alternative political-economy that might realize their own needs, wants, and desires in a far more egalitarian world.

Much has changed since the revolutionary times in which Marx and Engels penned their words. But, fortunately, the writers of the *Manifesto* freely acknowledged the contingency of its own making. 'The practical application of the principles,' wrote Marx and Engels (1952 edition, 8) in the 1872 Preface to the German edition, 'will depend, as the *Manifesto* itself states everywhere and at all times, on the historical conditions for the time being existing.' While we have not the right, they observe, to alter what has become a key historical document, we all have not only the right but also the obligation to interpret and re-charge it in the light of our own historical and geographical conditions. 'Does it require deep intuition,' they ask, 'to comprehend that man's ideas, views, and conceptions, in one word, man's consciousness, changes with every change in the conditions of his material existence, in his social relations and in his social life?' (72).

There are, of course, passages where the *Manifesto* appears quaint, outdated or downright objectionable to those of us who cling to socialist sentiments in these equally troubling but by no means revolutionary times. But there is much that comes through with such force and clarity that it is stunning to contemplate its contemporary relevance.

Consider, for example, some of those familiar passages that still strike at the core of contemporary alienations and sensibilities, most particularly as these have evolved within the era of free-market liberalism over the last twenty years. The bourgeoisie, say Marx and Engels:

> has left remaining no other nexus between man and man than naked self-interest, than callous 'cash payment'. It has drowned the most heavenly ecstasies of religious fervour, of chivalrous enthusiasm, of philistine sentimentalism, in the icy water of egotistical calculation. It has resolved personal worth into exchange value, and in place of the numberless inde-feasible chartered freedoms, has set up that single unconscionable freedom – Free Trade ... The bourgeoisie has stripped of its halo every occupation hitherto honoured and looked up to with reverent awe. It has converted the physician, the lawyer, the priest, the poet, the man of science, into its paid wage laborers.   (44)

Does this not describe with deadly accuracy the appalling powers that corrupt contemporary education, politics, social affairs, and moral senti-ments to such a degree that we are left with few options except to react against them by way of religious fundamentalism, mysticism, personal narcissism, and self-alienation? Are we not surrounded by the 'icy waters' of 'egotistical calculation' at every turn? And then consider this:

> The bourgeoisie cannot exist without constantly revolutionising the instru-ments of production, and thereby the relations of production, and with them the whole relations of society ... Constant revolutionising of produc-tion, uninterrupted disturbance of all social conditions, everlasting uncer-tainty and agitation distinguish the bourgeois epoch from all earlier ones. All fixed, fast-frozen relations, with their train of ancient and venerable prejudices and opinions, are swept away, all new formed ones become antiquated before they can ossify. All that is solid melts into air, all that is holy is profaned ...   (45–6)

The rhetorical power of such passages, the certitude of enunciation, the acute combination of admiration and horror for the immense powers unleashed under free-market capitalism (later compared, in one of those most striking of Faustian metaphors, to a 'sorcerer, who is no longer able to control the powers of the nether world whom he has called up by his spells') is impressive indeed.

The *Manifesto* also warns us of the inevitability of the crises that period-ically shake society to its very foundations, crises of creative destruction that are characterized by the 'absurdity' of overproduction in the midst of innumerable pressing but unfulfilled social needs, of famine in the midst of abundance, of spiraling inequalities, and of the periodic destruction of the previously created productive forces with which the bourgeoisie sought to create a world in its own image. We learn of the massive technological changes that completely transform the surface of the earth and our relation to nature (the 'subjection of Nature's forces to man, machinery, application of chemistry to industry and agriculture, steam navigation, railways, electric telegraphs, clearing whole continents for cultivation, canalisation of rivers, whole populations conjured out of the ground' [48]). But these same seemingly magical powers produce unemployment, disinvestment, and destruction of ways of life that even the bourgeoisie holds dear. So, how does the bourgeoisie get over these crises?

> On the one hand by enforced destruction of a mass of productive forces; on the other by the conquest of new markets, and by the more thorough exploitation of the old ones. That is to say, by paving the way for more extensive and more destructive crises, and by diminishing the means whereby crises are prevented. (50)

The crisis tendencies of capitalism widen and deepen at every turn.

The perceptive geographer will immediately detect the specifically spatial and geographical dimension to this argument. A closer inspection of the *Manifesto* reveals it to contain a distinctive polemic as to the role of geographical transformations, of 'spatial fixes,' and of uneven geographical developments, in the long history of capitalist accumulation. This dimension to the *Manifesto* deserves closer scrutiny, since it has much to say about how the bourgeoisie both creates and destroys the geographical foundations – ecological, spatial, and cultural – of its own activities, building a world in its own image. This is the central contradiction I wish to focus on here even if, as will soon become apparent, it is neither feasible nor desirable to isolate any one theme in the *Manifesto* from the rest.

1 The spatial dimension

The accumulation of capital has always been a profoundly geographical affair. Without the possibilities inherent in geographical expansion, spatial reorganization, and uneven geographical development, capitalism would long ago have ceased to function as a political-economic system. This perpetual turning to what I have elsewhere (see Harvey, 1982) termed 'a spatial fix' to capitalism's internal contradictions (most notably registered as an overaccumulation of capital within a particular geographical area) coupled with the uneven insertion of different territories and social formations into the capitalist world market has created a global historical geography of capital accumulation whose character needs to be well understood. Such differentiations are more important today than they ever were and the *Manifesto*'s weaknesses as well as its strengths in its approach to them need to be confronted and addressed. How Marx and Engels conceptualized the problem also deserves scrutiny because it was here that a European-wide communist movement – with representatives from many countries – came together to try to define a common revolutionary agenda that would work in the midst of considerable geographical and cultural differentiation.

The approach that Marx and Engels took to the problem of uneven geographical development and the spatial fix is somewhat ambivalent. On the one hand, questions of urbanization, geographical transformation, and 'globalization' are given a prominent place in their argument, but on the other hand the potential ramifications of geographical restructurings tend to get lost in a rhetorical mode that in the last instance privileges time and history over space and geography.

The opening sentence of the *Manifesto* situates the argument in Europe and it is to that transnational entity and its working classes that its theses are primarily addressed. This reflects the fact that 'Communists of various nationalities' (French, German, Italian, Flemish, and Danish, as well as English, are the languages envisaged for publication of the document) were assembled in London to formulate a working-class program. The document is, therefore, Eurocentric rather than international.

But the importance of the global setting is not ignored. The revolutionary changes that brought the bourgeoisie to power were connected to 'the discovery of America, the rounding of the Cape' and the opening up of trade with the colonies and with the East Indian and Chinese markets. The rise of the bourgeoisie is, from the outset of the argument, intimately connected to its geographical activities and strategies on the world stage:

> Modern industry has established the world market, for which the discovery of America paved the way. This market has given an immense development to commerce, to navigation, to communication by land. This development has in turn reacted on the extension of industry; in proportion as industry, commerce, navigation, railways extended, in the same proportion the bourgeoisie developed, increased its capital, and pushed into the background every class handed down from the Middle Ages.
>
> (Marx and Engels, 1952 edition, 42–3)

By these geographical means the bourgeoisie by-passed, undermined from the outside, and subverted from the inside place-bound feudal powers. By these means also the bourgeoisie converted the state (with its military, organizational, and fiscal powers) into the executive of its own ambitions (44). And, once in power, the bourgeoisie continued to pursue its revolutionary mission in part via both internal and external geographical transformations. Internally, the creation of great cities and rapid urbanization bring the towns to rule over the country (simultaneously rescuing the latter from the 'idiocy' of rural life and reducing the peasantry to a subaltern class). Urbanization concentrates productive forces as well as labor power in space, transforming scattered populations and decentralized systems of property rights into massive concentrations of political and economic power that are eventually consolidated in the legal and military apparatus of the nation state. 'Nature's forces' are subjected to human control as transport and communications systems, territorial divisions of labor, and urban infrastructures are created as foundations for capital accumulation.

But the consequent concentration of the proletariat in factories and towns makes them aware of their common interests. On this basis they begin to build institutions, such as unions, to articulate their claims (53–5). Furthermore, the modern systems of communications put 'the workers of different localities in contact with each other' thus allowing 'the numerous local struggles, all of the same character' to be centralized into 'one national struggle between the classes.' This process, as it spreads across frontiers, strips the workers of 'every trace of national character,' for each and every one of them is subject to the unified rule of capital (58). The organization of working-class struggle concentrates and diffuses across space in a way that mirrors the actions of capital (see below).

Marx expands on this idea in a passage that is so famous that we are apt to skim over it rather than read and reflect upon it with the care it deserves:

> The need for a constantly expanding market chases the bourgeoisie over the whole surface of the globe. It must settle everywhere, establish connexions everywhere ... The bourgeoisie has through its exploitation of the world market given a cosmopolitan character to production and consumption in every country ... All old established national industries have been destroyed or are daily being destroyed. They are dislodged by new industries, whose introduction becomes a life and death question for all civilized nations, by industries that no longer work up indigenous raw material, but raw material drawn from the remotest zones; industries whose products are consumed, not only at home, but in every quarter of the globe. In place of the old wants, satisfied by the production of the country, we find new wants, requiring for their satisfaction the products of distant lands and climes. In place of the old local and national seclusion and self-sufficiency, we have intercourse in every direction, universal interdependence of nations. And as in material, so also in intellectual production. The intellectual creations of individual nations become common property. National one-sidedness and narrow-mindedness become more and more impossible, and from the numerous national and local literatures, there arises a world literature ...   (46–7)

If this is not a compelling description of 'globalization' as we now know it then it is hard to imagine what would be. But Marx and Engels add something:

> The bourgeoisie ... draws all, even the most barbarian nations into civil-
> ization. The cheap prices of its commodities are the heavy artillery with
> which it batters down all Chinese walls, with which it forces the barbarians'
> intensely obstinate hatred of foreigners to capitulate. It compels all nations
> on pain of extinction, to adopt the bourgeois mode of production; it compels
> them to introduce what it calls civilization into their midst, i.e. to become
> bourgeois themselves. In one word, it creates a world after its own image. (47)

The theme of the 'civilizing mission' of the bourgeoisie is here enunciated
(albeit with a touch of irony). But a certain limit to the power of
geographical expansion to work indefinitely and in perpetuity is implied.
If the geographical mission of the bourgeoisie is the reproduction of class
and productive relations on a progressively expanding geographical scale,
then the field of play for both the internal contradictions of capitalism and
for socialist revolution likewise expands geographically. Class struggle
becomes global which leads, of course, to the famous imperative 'working
men of all countries unite' as a necessary condition for an anti-capitalist
and pro-socialist revolution.

2 The theory of the spatial fix

Marx and Engels did not formulate their ideas out of the blue. They in fact
appealed to a long tradition of analysis. In particular, they seem to have
relied significantly upon a reading of Hegel's *The Philosophy of Right*,
drawing both strengths and weaknesses from the inspiration of that text.

Hegel (1967 edition, 148–52, 278) there presented imperialism and
colonialism as potential solutions to the serious and stressful internal
contradictions of what he considered to be a 'mature' civil society. The
increasing accumulation of wealth at one pole, and the formation of a
'penurious rabble' trapped in the depths of misery and despair at the
other, sets the stage for social instability and class war that cannot be cured
by any internal transformation (such as a redistribution of wealth from
rich to poor). Civil society is thereby driven by its 'inner dialectic' to 'push
beyond its own limits and seek markets, and so its necessary means of
subsistence, in other lands that are either deficient in the goods it has
overproduced, or else generally backward in industry.' It must also found
colonies and thereby permit a part of its population 'a return to life on the
family basis in a new land.' By this means also it 'supplies itself with a new
demand and field for its industry.' All of this is fueled by a 'passion for
gain' that inevitably involves risk so that industry, 'instead of remaining
rooted to the soil and the limited circle of civil life with its pleasures and
desires, ... embraces the element of flux, danger, and destruction.' A
prefiguring of some of the rhetoric in the *Manifesto* is already evident.

Having, in a few brief startling paragraphs, sketched the possibilities of an imperialist and colonial solution to the ever-intensifying internal contradictions of civil society, Hegel rather surprisingly drops the matter. He leaves us in the dark as to whether capitalism could be stabilized by appeal to some sort of 'spatial fix' in either the short or long run. Instead, he turns his attention to the concept of the state as the actuality of the ethical idea. This could be taken to imply that transcendence of civil society's internal contradictions by the modern state – an inner transformation – is both possible and desirable. Yet Hegel nowhere explains how the problems of poverty and of the increasing polarization in the distribution of wealth he has already identified are actually to be overcome. Are we supposed to believe, then, that these particular problems can be dealt with by imperialism? The text is ambivalent. This is, as Avineri (1972, 132) points out, 'the only time in his system, where Hegel raises a problem – and leaves it open.' There is, it seems, just a possibility that a solution to capitalism's problem lies in some promised land or other space beyond the horizon.

How far Hegel influenced Marx's later concerns can be endlessly debated. Engels certainly believed that Marx was 'the only one who could undertake the work of extracting from Hegelian logic the kernel containing Hegel's real discoveries' (Marx and Engels, 1980 edition, 474). The language Marx uses to describe the general law of capitalist accumulation in *Capital,* for example, bears a close resemblance to that of Hegel. *The Philosophy of Right* (Hegel, 1967 edition, 150) states:

> When the standard of living of a large mass of people falls below a certain subsistence level – a level regulated automatically as the one necessary for a member of the society . . . the result is the creation of a rabble of paupers. At the same time this brings with it, at the other end of the social scale, conditions which greatly facilitate the concentration of disproportionate wealth in a few hands.

In *Capital*, Volume 1, Marx writes:

> [A]s capital accumulates, the situation of the worker, be his payment high or low, must grow worse . . . It makes an accumulation of misery a necessary condition, corresponding to the accumulation of wealth. Accumulation of wealth at one pole is, therefore, at the same time accumulation of misery, the torment of labour, slavery, ignorance, brutalization and moral degradation at the opposite pole, i.e. on the side of the class that produces its own product as capital. (799)

The parallel between the two texts is striking. It is even possible to interpret the first volume of *Capital* as a tightly orchestrated argument, buttressed by a good deal of historical and material evidence, to prove that the propositions Hegel had so casually advanced, without any logical or

evidentiary backing, were indubitably correct. The internal contradictions
that Hegel depicted were, in Marx's view, not only inevitable but also
incapable of any internal resolution short of proletarian revolution. Left to
its own devices, unchecked and unregulated, free-market capitalism would
end up depleting and ultimately destroying the two sources of its own
wealth – the laborer and the soil. This was the conclusion that Marx
wanted to force not only upon the Hegelians but upon everyone else. But
in order to make the argument stick he also had to bear in mind the
question of some spatial fix – some utopian solution in some other space –
that Hegel had raised but left open.

In this light one other feature in the structure of argument in *Capital*
makes sense. The last chapter of Volume 1 deals with the question of
colonization. It seems, at first sight, an odd afterthought to a work which,
in the preceding chapter announced expropriation of the expropriators and
the deathknell of the bourgeoisie with a rhetoric reminiscent of the
Manifesto. But in the light of Hegel's argument the chapter acquires a
particular significance. Marx seeks to show how the bourgeoisie contra-
dicted its own myths as to the origin and nature of capital by the policies it
advocated in the colonies. In bourgeois accounts (the paradigmatic case
being that of Locke), capital (a thing) originated in the fruitful exercise of
the producer's own capacity to labor, while labor power as a commodity
arose through a social contract, freely entered into, between those who
produced surplus capital through frugality and diligence, and those who
chose not to do so. 'This pretty fancy,' Marx thunders, is 'torn asunder' in
the colonies. As long as the laborer can 'accumulate for himself – and this
he can do as long as he remains possessor of his means of production –
capitalist accumulation and the capitalist mode of production are impossible.' Capital is not a physical thing but a social relation. It rests on the
'annihilation of self-earned private property, in other words, the expropriation of the laborer.' Historically, this expropriation was 'written in the
annals of mankind in letters of blood and fire' – and Marx cites chapter,
verse, and the Duchess of Sutherland to prove his point. This same truth is
expressed in colonial land policies, such as those of Wakefield in Australia,
in which the powers of private property and the state were to be used to
exclude laborers from easy access to free land in order to preserve a pool of
wage laborers for capitalist exploitation. Thus was the bourgeoisie forced
to acknowledge in its program of colonization what it sought to conceal at
home – that wage labor and capital are both based on the forcible separation of the laborer from control over the means of production. This is the
secret of 'primitive' or 'original' capital accumulation.

The relation of all this to the question Hegel left open needs explication. If laborers can return to a genuinely unalienated existence (establish

their utopia) through migration overseas or to some frontier region, then capitalist control over labor supply is undermined. Such a form of expansion may be advantageous to labor but it could provide no solution to the inner contradictions of capitalism. In later texts Marx brought the issue into sharper focus. He first distinguished between two types of colonial venture:

> There are colonies proper, such as the United States, Australia, etc. Here the mass of the farming colonists, although they bring with them a larger or smaller amount of capital from the motherland, are not capitalists, nor do they carry on capitalist production. They are more or less peasants who work for themselves and whose main object, in the first place, is to produce their own livelihood . . . In the second type of colonies – plantations – where commercial speculations figure from the start and production is intended for the world market, the capitalist mode of production exists.
>
> (Marx, 1968 edition, 302–3)

In the first kind of colony:

> [T]he capitalist regime everywhere comes into collision with the resistance of the producer, who, as owner of his own conditions of labour, employs that labour to enrich himself, instead of the capitalist. The contradiction of these two diametrically opposed economic systems, manifests itself here practically in a struggle between them. Where the capitalist has at his back the power of the mother country, he tries to clear out of his way by force, the modes of production and appropriation, based on the independent labour of the producer.   (Marx, 1976 edition, 716)

The new markets and new fields for industry which Hegel saw as vital could be achieved only through the re-creation of capitalist relations of private property and the associated power to appropriate the labor of others. The fundamental conditions which gave rise to the problem in the first place – alienation of labor – are thereby replicated. Marx's chapter on colonization appears to close off the possibility of any external 'spatial fix' to the internal contradictions of capitalism. Marx evidently felt obliged to close the door that Hegel had left partially ajar and consolidate his call for total revolution by denying that colonization (or by extension any other kind of spatial fix) could, in the long run, be a viable solution to the inner contradictions of capitalism.

But the door will not stay shut. Hegel's 'inner dialectic' undergoes successive representations in Marx's work and at each point the question of the spatial resolution to capitalism's contradictions can legitimately be posed anew. The chapter on colonization may suffice for the first volume of *Capital* where Marx concentrates solely on questions of production. But what of the third volume where Marx shows that the requirements of production conflict with those of circulation to produce crises of over-accumulation? Polarization then takes the form of 'unemployed capital at one pole and unemployed worker population at the other' (Marx, 1967 edition, Volume 3, 251) and the consequent devaluation of both. Can the formation of such crises be contained through geographical expansions and restructurings? Marx, as I have shown elsewhere (Harvey, 1982), does not rule out the possibility that foreign trade and growth of external markets, the export of capital for production, and the expansion of the proletariat through primitive accumulation in other lands, can counteract the falling rate of profit in the short run. But how long is the short run? And if it extends over many generations (as Rosa Luxemburg in her theory of imperialism implied), then what does this do to Marx's theory and its associated political practice of seeking for revolutionary transformations in the heart of civil society in the here and now?

And what if the workers also sought their own spatial fix? Marx is then obliged to confront a growing belief within the workers' movement that escape to some promised land through emigration was the answer to their ills. Marx's open letter (cited in Marin, 1984) to the Icarians, a utopian sect led by the influential Etienne Cabet, written in 1848, not only argues that any new foundation for social organization must be laid in place, in Europe, but that the attempt to escape to Utopia (no place – Icaria) was doomed to fail. The emigrants, he argued, would be 'too infected with the errors of their education and prejudices of today's society to be able to get rid of them in Icarie.' Internal dissensions would easily be exploited by hostile and alien external powers. Furthermore, the burdens placed on workers through radical changes in divisions of labor and environmental conditions would demand a level of enthusiasm and commitment that was bound to wane with time. And, Marx goes on to observe perceptively, 'a few hundred thousand people cannot establish and continue a communal living situation without it taking on an absolutely exclusive and sectarian nature.' These turned out, of course, to be exactly the problems that led to the demise of Icarian settlements in the United States (see Johnson, 1974), thus giving some credibility to Marx's closing argument:

> For communists – and surely Icarians – who realize the principle of per-sonal freedom, a community of communal property without a transition period, actually a democratic transition where personal property is slowly transformed into social property, is as impossible as is harvesting grain without having planted.   (Marin, 1984, 273–9)

For the workers, utopian longings may have been understandable, but there was no real way that the spatial fix could work for them any more than it could for capital.

3 Problematizing the *Manifesto*'s geography

The geographical element in the *Manifesto* has to a large degree been ignored in subsequent commentaries. When it has been the focus of attention it has often been treated as unproblematic in relation to political action. This suggests a twofold response as we look back upon the argument. First, it is vital to recognize (as the *Manifesto* so clearly does) the ways in which geographical reorderings and restructurings, spatial strategies and geopolitical elements, uneven geographical developments, and the like, are vital aspects to the accumulation of capital and the dynamics of class struggle, both historically and today. It is likewise vital to recognize (in ways the *Manifesto* tends to underplay) that class struggle unfolds differentially across this highly variegated terrain and that the drive for socialism must take geographical realities and geopolitical possibilities into account.

But, secondly, it is equally important to problematize the actual account ('sketch' might be a more appropriate word) given in the *Manifesto* in order to develop a more sophisticated, accurate, and politically useful understanding as to how the geographical dimensions to capital accumulation and class struggle have played and continue to play such a fundamental role in the perpetuation of bourgeois power and the suppression of worker rights and aspirations.

In what follows, I shall largely take the first response as a 'given' even though I am only too aware that it needs again and again to be reasserted within a movement that has not by any means taken on board some, let alone all, of its basic implications. While Lefebvre (1976) perhaps exaggerates a touch, I think it worth recalling his remark that capitalism has survived in the twentieth century by one and only one means – 'by occupying space, by producing space.' It would be ironic indeed if the same were to be said at the end of the twenty-first century!

My main concern here, then, is to fashion a critical evaluation of the actual account given in the *Manifesto*. I do so mainly from the standpoint of our own times rather than from the perspective of 1848 (although, as I shall occasionally indicate, there are some points where the *Manifesto* is open to challenge even at its own moment of conception). I shall, in the process, try to detach the underpinnings of the argument concerning the spatio-temporal development of capital accumulation and class struggle from its Hegelian basis. From such a perspective I isolate seven aspects of the *Manifesto*'s geography for critical commentary.

1. The division of the world into 'civilized' and 'barbarian' nations is, to say the least, anachronistic if not downright objectionable even if it can be excused as typical of the times. Furthermore, the generalized

center-periphery model of capital accumulation which accompanies it is at best a gross oversimplification and at worst misleading. It makes it appear as if capital centered in one place (England or Europe) diffuses outwards to encompass the rest of the world. This idea seems to derive from uncritical acceptance of Hegel's teleology – if space is to be considered at all it is as a passive recipient of a teleological historical process that starts from the center and flows outwards to fill up the entire globe. Leaving aside the whole problem of where, exactly, capitalism was born and whether it arose in one and only one place or was simultaneously emerging in geographically distinctive environments – an arena of scholarly dispute that shows no sign of coming to a consensus (see Blaut, 1977; 1993) – the subsequent development of a capitalism that had, by the end of the eighteenth century at least, come to concentrate its freest forms of development in Europe in general and Britain in particular, cannot be encompassed by such a diffusionist way of thinking. While there are some instances in which capital diffused outwards from a center to a periphery (e.g. the export of surplus capital from Europe to Argentina or Australia in the late nineteenth century), such an account is inconsistent with what happened in Japan after the Meiji restoration or what is happening today as first South Korea and then China engages in some form of internalized primitive accumulation and inserts its labor power and its products into global markets.

The geography of capital accumulation deserves a far more nuanced treatment than the diffusionist sketch provided in the *Manifesto*. The problem does not lie in the sketchiness of the account *per se*, but in the failure to delineate a theory of uneven geographical development (often entailing uneven primitive accumulation) that would be helpful for charting the dynamics of working-class formation and class struggle across even the European, let alone the global, space. Marx partially rectified this problem in later works. And there is at least a hint in the *Manifesto* of a more dialectical reading of the origins of capital in the mercantile activities of appropriation and plundering of wealth from around the world.

A more fully theorized understanding of the space/place dialectic in capitalist development would also be helpful. How do places, regions, territories evolve given changing space relations? Geopolitical games of power between nation states (or other territorial units), for example, become interconnected with market position in a changing structure of space relations which, in turn, privileges certain locations and territories for capitalist accumulation. It is also interesting to note how those national bourgeoisies that could not easily use spatial powers to circumvent feudalism ended up with fascism (twentieth-century Germany,

Italy, and Spain are cases in point). Since these are rather abstract arguments, I shall try to put some flesh and bones on them in what follows.

To begin with, the globe never has been a level playing field upon which capital accumulation could play out its destiny. It was and continues to be an intensely variegated surface, ecologically, politically, socially, and culturally differentiated. Flows of capital found some terrains easier to occupy than others in different phases of development. And in the encounter with the capitalist world market, some social formations adapted to insert themselves aggressively into capitalistic forms of market exchange while others did not, for a wide range of reasons and with consummately important effects. Primitive or 'original' accumulation on the part of some non-capitalistic ruling class can and has occurred in different places and times, albeit facilitated by contact with the market network that increasingly pins the globe together into an economic unity. But how and where that primitive accumulation occurs depends upon local conditions even if the effects are global. It is now a widely held belief in Japan, for example, that the commercial success of that country after 1960 was in part due to the non-competitive and withdrawn stance of China after the revolution and that the contemporary insertion of Chinese power into the capitalist world market spells doom for Japan as a producer as opposed to a *rentier* economy. Contingency of this sort rather than teleology has a lot of play within capitalist world history. Furthermore, the globality of capital accumulation poses the problem of a dispersed bourgeois power (and complicated relations and alliances with non-capitalist ruling elites) that can become much harder to handle geopolitically precisely because of its multiple sites. Marx himself later worried about this political possibility. In 1858 he wrote, in a passage that Meszaros (1995) rightly makes much of:

> For us the difficult question is this: the revolution on the Continent is imminent and its character will be at once socialist; will it not be *necessarily crushed* in this *little corner of the world*, since on a much larger terrain the development of bourgeois society is still *in the ascendant*.   (XII)

It is chastening to reflect upon the number of socialist revolutions around the world that have been successfully encircled and crushed by the geopolitical strategies of an ascendant bourgeois power.

2. The *Manifesto* correctly highlights the importance of reducing spatial barriers through innovations and investments in transport and communications as critical to the growth and sustenance of bourgeois power. Moreover, the argument indicates that this is an on-going

rather than already-accomplished process. In this respect the *Manifesto* is prescient in the extreme. 'The annihilation of space through time,' as Marx later dubbed it, is deeply embedded in the logic of capital accumulation, entailing as it does the continuous though often lumpy transformations in space relations that have characterized the historical-geography of the bourgeois era (from turnpikes through railroads, highways, and air travel to cyberspace). These transformations undercut the absolute qualities of space (often associated with feudalism) and emphasize the relativity of space relations and locational advantages thus making the Ricardian doctrine of comparative advantage in trade a highly dynamic rather than stable affair. Furthermore, spatial tracks of commodity flows have to be mapped in relation to flows of capital, labor power, military advantage, technology transfers, information flows, and the like. In this regard the *Manifesto* was not wrong but underappreciated for its prescient statements.

3. Perhaps one of the biggest absences in the *Manifesto* is its lack of attention to the territorial organization of the world in general and of capitalism in particular. If, for example, the state was necessary as an 'executive arm of the bourgeoisie' then the state had to be territorially defined, organized, and administered. The sketch of how this occurred is provocative but far too brief. The concentration of property and the rise of the bourgeois class to political ascendancy 'lumped together' independent or 'loosely connected provinces with separate interests, laws, governments and systems of taxation' into 'one nation, with one government, one code of laws, one national class interest, one frontier and one customs tariff' (Marx and Engels, 1952 edition, 48).

While the right of sovereign independent states to co-existence was established at the Treaty of Westphalia in 1648 as a (distinctively shaky) European norm, the general extension of that principle across the globe took several centuries to take shape as did the internal processes of nation state formation. In 1848 the 'lumping together' had yet to occur in Germany and Italy. In much of the rest of the world it is arguably even now not yet accomplished. The nineteenth century was the great period of territorial definitions (with most of the world's boundaries being established between 1870 and 1925 and most of those being drawn by the British and the French alone – the carve-up of Africa in 1885 being the most spectacular example). But state formation and consolidation is quite another step beyond territorial definition and it has proven a long drawn out and often unstable affair (particularly, for example, in Africa). Only after 1945 did decolonization push state formation worldwide a bit closer to the highly simplified model that the *Manifesto* envisages. Furthermore, the relativism introduced by revolutions in transport and communications coupled with the uneven dynamics of class struggle and uneven resource endowments means that territorial configurations cannot remain stable for long. Flows of commodities, capital, labor, and information always render boundaries porous. There is plenty of play for contingency here (including phases of territorial reorganization and redefinition), thus undermining any simple teleological interpretation (of the sort that derives from Hegel and which can still be found in both capitalistic and communist ideas about what the future necessarily holds).

4. The state is, of course, only one of many mediating institutions that influence the dynamics of accumulation and class struggle. Money and finance must also be given pride of place. In this respect there are some intriguing questions about which the *Manifesto* remains silent, in part, I suspect, because its authors had yet to discover their fundamental insights about the dialectical relations between money, production, commodity exchange, distribution, and production (as these are conceptualized, for example, in the Introduction to the *Grundrisse*). But it may also have been that the authors were ambivalent in the face of the two main strains of radical thinking at the time – that of the Saint-Simonians who looked to the association of capitals and the centralization of credit as a solution versus what was to become the anarchist stress (of, e.g., Proudhon) upon decentralization and credit cooperatives with its strong appeal to artisans and petty commodity producers. There are two ways to look at this (and I here take the question of money and credit as both emblematic and fundamental). On the one hand we can interpret world money as some universal representation of value to which territories relate (through their own currencies) and to which capitalist producers conform as they seek some measure of their performance and profitability. This is a very functionalist and undialectical view. It makes it seem as if value hovers as some ethereal abstraction over the activities of individuals as of nations (this is, incidentally, the dominant conception at work in the contemporary neoclassical ideology of globalization). In *Capital*, Marx looks upon world money differently, as a representation of value that arises out of a dialectical relation between the particularity of material activities (concrete labor) undertaken in particular places and times and the universality of values (abstract labor) achieved as commodity exchange becomes so widespread and generalized as to be a normal social act. But institutions mediate between particularity and universality so as to give some semblance of order and permanence to what is otherwise as shifting sand. Central banks, financial institutions, exchange systems, state-backed local currencies, etc., then become powerful mediators

between the universality of money on the world market and the particularities of concrete labors conducted here and now around us. Such mediating institutions are also subject to change as, for example, powers shift from yen to deutsch marks to dollars and back again or as new institutions (like the IMF and the World Bank after 1945) spring up to take on new mediating roles.

The point here is that there is always a problematic relation between local and particular conditions on the one hand and the universality of values achieved on the world market on the other and that this internal relation is mediated by institutional structures which themselves acquire a certain kind of independent power. These mediating institutions are often territorially based and biassed in important ways. They play a key role in determining where certain kinds of concrete labors and class relations shall arise and can sometimes even dictate patterns of uneven geographical development through their command over capital assembly and capital flows. Given the importance of European-wide banking and finance in the 1840s (the Rothschilds being prominent players in the events of 1848) and the political-economic theories of the Saint-Simonians with respect to the power of associated capitals to change the world, the absence of any analysis of the mediating institutions of money and finance is surprising even if one of the key political proposals in the *Manifesto* is 'centralisation of credit in the hands of the state, by means of a national bank with state capital and an exclusive monopoly' (75). Subsequent formulations (not only by Marx but also by Lenin, Hilferding, and many others – see Harvey [1982] for a summary) may have helped to elaborate on such matters, but the rather episodic and contingent treatment of the role of finance and money capital in organizing the geographical dynamics of capital accumulation may have been one of the *Manifesto*'s unwitting and unfortunate legacies (hardly anything was written on the topic between Hilferding's guiding work of 1910 and the early 1970s).

5. The argument that the bourgeois revolution subjugated the countryside to the city, that processes of industrialization and rapid urbanization laid the seedbed for a more united working-class politics, is of major importance. Reduced to its simplest formulation, it says that the production of spatial organization is not neutral with respect to class struggle. And that is a vital principle no matter how critical we might be with respect to the sketch of these dynamics as laid out in the *Manifesto*. The account offered runs like this:

> The proletariat goes through various stages of development. With its birth begins its struggle with the bourgeoisie. At first the contest is

carried on by individual labourers, then by the workpeople of a factory, then by the operatives of one trade, in one locality, against the individual bourgeois who directly exploits them. At this stage the labourers still form an incoherent mass scattered over the country, and broken up by their mutual competition. If anywhere they unite to form more compact bodies, this is not yet the consequence of their own active union, but of the union of the bourgeoisie . . . But with the development of industry the proletariat not only increases in number; it becomes concentrated in greater masses, its strength grows, and it feels that strength more . . . the collisions between individual workmen and individual bourgeois take more and more the character of collisions between two classes. There-upon the workers begin to form combinations (Trades' Unions) . . . This union [of the workers] is helped on by the improved means of commu-nication that are created by modern industry and that place the workers of different localities in contact with one another. It was just this contact that was needed to centralise the numerous local struggles, all of the same character, into one national struggle between classes . . .   (54–5)

For much of the nineteenth century this account captures a common enough path to the development of class struggle. And there are plenty of twentieth-century examples where similar trajectories can be discerned (the industrialization of South Korea being paradig-matic). But it is one thing to say that this is a useful descriptive sketch and quite another to argue that these are necessary stages through which class struggle must evolve *en route* to the construction of socialism.

If, furthermore, it is interpreted, as I have suggested, as a compel-ling statement of the non-neutrality of spatial organization in the dynamics of class struggle, then it follows that the bourgeoisie may also evolve its own spatial strategies of dispersal, of divide and rule, of geographical disruptions to the rise of class forces that so clearly threaten its existence. To the passages already cited we find added the cautionary statement that: 'this organization of the proletarians into a class, and consequently into a political party, is continually being upset again by the competition between the workers themselves' (55). And there are plenty of examples of bourgeois strategies to achieve that effect. From the dispersal of manufacturing from centers to suburbs in late-nineteenth-century US cities to avoid concentrated proletarian power to the current attack on union power by dispersal and fragmentation of production processes across space (much of it, of course, to so-called developing countries where working-class organization is weakest) has proven a powerful weapon in the bourgeois struggle to enhance its power. The active stimulation of inter-worker competition across space has likewise worked to capitalist advantage to say nothing of the problem of localism and nationalism within working-class movements (the position of the Second International in the First World War being the most spectacular case). In general, I think it fair to say that workers' movements have been better at commanding power in places and territories rather than in controlling spatialities with the result that the capitalist class has used its superior powers of spatial maneuver to defeat place-bound proletarian/socialist revolutions (cf. Marx's 1858 worry cited above). The recent geographical and ideological assault on working-class forms of power through 'globalization' gives strong support to this thesis (see Chapter 3). While none of this is inconsistent with the basic underpinning of the argument in the *Manifesto*, it is, of course, different from the actual sketch of class-struggle dynamics set out as a stage model for the development of socialism in the European context of 1848.

6. The general presumption throughout the *Manifesto* is that the nexus for revolutionary action lies mainly with a rapidly urbanizing industrial proletariat. Even at the time, ignoring the revolutionary potential of rural, agricultural, and peasant-based movements must have appeared somewhat premature (the Tolpuddle Martyrs were agricultural workers who tried to form a union in Dorset only to be exiled to Australia for their pains, and many segments of the French countryside were as alive in 1848 with the same revolutionary sentiments that had made them such key players in the French Revolution). The long subsequent history of peasant struggles and guerilla wars, to say nothing of struggles waged by petty commodity producers, plantation workers, and other laborers in agriculture, have brought into question the *Manifesto*'s central presumption of where the potential for revolutionary action (and reaction) was to be found.

7. This leads us to one of the most problematic elements in the *Manifesto*'s legacy: the homogenization of the 'working man' and of 'labor powers' across a highly variegated geographical terrain as the proper basis for struggles against the powers of capital. While the slogan 'working men of all countries unite' may still stand (suitably modified to rid it of its gendered presupposition) as the only appropriate response to the globalizing strategies of capital accumulation, the manner of arriving at and conceptualizing that response deserves critical scrutiny. Central to the argument lies the belief that modern industry and wage labor, imposed by the capitalists ('the same in England as in France, in America as in Germany'), have stripped the workers 'of every trace of national character.' As a result:

The working men have no country. We cannot take from them what they have not got. Since the proletariat must first of all acquire political supremacy, must rise to be the leading class of the nation, must constitute itself the nation, it is, so far, itself national, though not in the bourgeois sense of the word.

National differences and antagonisms between peoples are daily more and more vanishing, owing to the development of the bourgeoisie, to freedom of commerce, to the world market, to uniformity in the mode of production and in the conditions of life corresponding thereto.

The supremacy of the proletariat will cause them to vanish still faster. United action, of the leading civilised countries at least, is one of the first conditions for the emancipation of the proletariat.

In proportion as the exploitation of one individual by another is put an end to, the exploitation of one nation by another will also be put an end to. In proportion as the antagonism between classes within the nation vanishes, the hostility of one nation to another will come to an end.

(71–2)

The guiding vision is noble enough but there is unquestionably a lot of wishful thinking here. At best, the *Manifesto* mildly concedes that the initial measures to be taken as socialists come to power will 'of course be different in different countries.' It also notes how problems arise in the translation of political ideas from one context to another – the Germans took on French ideas and adapted them to their own circumstances which were not so well developed, creating a German kind of socialism that Marx was highly critical of (82–3). In the practical world of politics, then, there is a certain sensitivity to uneven material conditions and local circumstances. And in the final section of the *Manifesto* (94–6) attention is paid to the different political conditions in France, Switzerland, Poland, and Germany. From this Marx and Engels divine that the task of communists is to bring unity to these causes, to define the commonalities within the differences, and to make a movement in which workers of the world can unite. But in so doing the force of capital that uproots and destroys local place-bound loyalties and bonds is heavily relied upon to prepare the way.

There are, I think, two ways in which we can read this in relation to our contemporary conditions. On the one hand the *Manifesto* insists, correctly in my view, that the only way to resist capitalism and transform society towards socialism is through a global struggle in which global working-class formation, perhaps achieved in a step-wise fashion from local to national to global concerns, acquires sufficient power and presence to fulfill its own historical potentialities. In this case, the task of the communist movement is to find ways, against all odds, to bring together all the various highly differentiated and often local movements into some kind of commonality of purpose (cf. Moody, 1997; Herod, 1997; 1998). The second reading is rather more mechanistic. It sees the automatic sweeping away of national differences and differentiations through bourgeois advancement, the de-localization and de-nationalization of working-class populations and therefore of their political aspirations and movements. The task of the communist movement is to prepare for and hasten on the end point of this bourgeois revolution, to educate the working class as to the true nature of their situation and to organize, on that basis, their revolutionary potential to construct an alternative. Such a mechanistic reading is, in my view, incorrect, even though substantial grounding for it can be found within the *Manifesto* itself.

The central difficulty lies in the presumption that capitalist industry and commodification will lead to homogenization of the working population. There is, of course, an undeniable sense in which this is true, but what it fails to appreciate is the way in which capitalism simultaneously differentiates among workers, sometimes feeding off ancient cultural distinctions, gender relations, ethnic predilections, and religious beliefs. It does this not only through the development of explicit bourgeois strategies of divide and control, but also by converting the principle of market choice into a mechanism for group differentiation. The result is the implantation of all manner of class, gender, and other social divisions into the geographical landscape of capitalism. Divisions such as those between cities and suburbs, between regions as well as between nations cannot be understood as residuals from some ancient order. They are not automatically swept away. They are actively produced through the differentiating powers of capital accumulation and market structures. Place-bound loyalties proliferate and in some respects strengthen rather than disintegrate through the mechanisms of class struggle as well as through the agency of both capital and labor working for themselves. Class struggle all too easily dissolves into a whole series of geographically fragmented communitarian interests, easily coopted by bourgeois powers or exploited by the mechanisms of neo-liberal market penetration.

There is a potentially dangerous underestimation within the *Manifesto* of the powers of capital to fragment, divide, and differentiate, to absorb, transform, and even exacerbate ancient cultural divisions, to produce spatial differentiations, to mobilize geopolitically, within the overall homogenization achieved through wage labor and market exchange. And there is likewise an underestimation of the ways in which labor mobilizes through territorial forms of organization, building place-bound loyalties *en route*. The dialectic of commonality and difference has not worked out (if it ever could) in the way that the sketch supplied in the *Manifesto* implied, even if its underlying logic and its injunction to unite is correct.

CHAPTER 3

'Working Men of All Countries, Unite!'

Even though the *Manifesto* mainly concentrates on the conquest of nation-state power as its central political objective, the geographical logic of its argument also points towards a grander goal expressed in the final exhortation for workers of all countries to unite in anti-capitalist struggle. The conditions of global working-class formation in the last half of the twentieth century suggests that such an exhortation is more important than ever.

The World Bank (1995) estimates that the global labor force doubled in size between 1966 and 1995. By the latter date, an estimated 2.5 billion men and women were active participants in labor markets and thereby captive to the conditions of wage labor. Most of this wage labor force was living under the most appalling conditions. The World Bank report continues:

> [T]he more than a billion individuals living on a dollar or less a day depend ... on pitifully low returns to hard work. In many countries workers lack representation and work in unhealthy, dangerous, or demeaning conditions. Meanwhile 120 millions or so are unemployed worldwide, and millions more have given up hope of finding work.   (1–2)

This condition exists at a time of rapid growth in average levels of productivity per worker (reported also to have doubled since 1965 world-wide) and a rapid growth in world trade fueled in part by reductions in costs of movement but also by a wave of trade liberalization and sharp increases in the international flows of direct investments. The latter helped construct transnationally integrated production systems largely organized through intra-firm trade. As a result, says a report of the International Labour Office (1996):

> [T]he number of workers employed in export- and import-competing industries has grown significantly. In this sense, therefore, it could be said that labour markets across the world are becoming more interlinked ... Some observers see in these developments the emergence of a global labour

41

market wherein 'the world has become a huge bazaar with nations peddling their workforces in competition against one another, offering the lowest prices for doing business' ... The core apprehension is that intensifying global competition will generate pressures to lower wages and labour standards across the world.   (4)

This process of ever-stronger interlinkage has been intensified by 'the increasing participation in the world economy of populous developing countries such as China, India and Indonesia' (4). With respect to China, for example, the United Nations Development Program (1996) reports:

> The share of labour-intensive manufactures in total exports rose from 36% in 1975 to 74% in 1990 ... Between 1985 and 1993 employment in textiles increased by 20%, in clothing and fibre products by 43%, in plastic products by 51%. China is now a major exporter of labour-intensive products to many industrial countries ... For all its dynamic job creation, China still faces a formidable employment challenge. Economic reforms have released a 'floating population' of around 80 million most of whom are seeking work. The State Planning Commission estimates that some 20 million workers will be shed from state enterprises over the next five years and that 120 million more will leave rural areas hoping for work in the cities. Labour intensive economic growth will need to continue at a rapid pace if all these people are to find work.   (94)

I quote this instance to illustrate the huge movements into the global labor force that have been and are underway. And China is not alone in this. The export-oriented garment industry of Bangladesh hardly existed twenty years ago, but it now employs more than a million workers (80 per cent of them women and half of them crowded into Dhaka). Cities like Jakarta, Bangkok, and Bombay, as Seabrook (1996) reports, have become meccas for formation of a transnational working class – heavily dependent upon women – living under conditions of poverty, violence, chronic environmental degradation, and fierce repression.

It is hardly surprising that the insertion of this proletarianized mass into global trading networks has been associated with wide-ranging social convulsions and upheavals (see, e.g., some of the excellent documentation provided by Moody, 1997) as well as changing structural conditions, such as the spiraling inequalities between regions (that left sub-Saharan Africa far behind as East and South-East Asia surged ahead) and between classes. As regards the latter, *The UN Development Report* (1996) states that 'between 1960 and 1991 the share of the richest 20% rose from 70% of global income to 85% – while that of the poorest declined from 2.3% to 1.4%.' By 1991, 'more than 85% of the world's population received only 15% of its income' and 'the net worth of the 358 richest people, the dollar billionaires, is equal to the combined income of the poorest 45% of the world population – 2.3 billion people' (2). In the United States, the net wealth of Bill Gates alone in 1995 was greater than the combined net worth of the poorest 40 percent of Americans (106 million people).

This polarization is astounding, rendering hollow the World Bank's extraordinary claim that international integration coupled with free-market liberalism and low levels of government interference is the best way to deliver growth and to raise living standards for workers. (While growth was strong they oddly attributed these virtues to the economies of East and South-East Asia but when problems developed in the region the Bank and IMF changed their rhetoric to complain of too much state involvement.) It also renders hollow a wide-ranging set of ideological claims that the free market will create a 'stakeholder' society within a rapidly democratizing capitalism. Within the era of the 'so-called people's market' in the United States, for example, 'Federal Reserve statistics show that 60 percent of Americans own no stock at all' (not even through pension funds). Furthermore, 'the wealthiest 1 percent of Americans own nearly 50 percent of all stock; the bottom 80 percent own only 3 percent' (Smith, 1998, B18). It is then not hard to see who has benefitted most from the run-up in the stock market over the last decade. 'You are horrified at our intending to do away with private property,' Marx and Engels exclaim in the *Manifesto*, 'but in your existing society, private property is already done away with for nine-tenths of the population' (65).

The local reality of global inequality is described vividly by Seabrook (1996):

> Indonesia, in the name of the free market system, promotes the grossest violations of human rights, and undermines the right to subsist of those on whose labour its competitive advantage rests. The small and medium-sized units which subcontract to the multinationals are the precise localities where the sound of the hammering, tapping, beating of metal comes from the forges where the chains are made for industrial bondage . . .
>
> Many transnationals are subcontracting here: Levi Strauss, Nike, Reebok. A lot of the subcontractors are Korean-owned. They all tend to low wages and brutal management. Nike and Levis issue a code of conduct as to criteria for investment; but in reality, under the tender system they always go for the lowest cost of production . . . Some subcontractors move out of Jakarta to smaller towns, where workers are even less capable of combining to improve their conditions. (103–5)

Or, at a more personal level there is the account given by two sisters, Hira and Mira, who until recently worked for a Singaporean-owned subcontractor for Lévi Strauss:

'We are regularly insulted, as a matter of course. When the boss gets angry
he calls the women dogs, pigs, sluts, all of which we have to endure
patiently without reacting . . . We work officially from seven in the morning
until three (salary less than $2 per day), but there is often compulsory
overtime, sometimes – especially if there is an urgent order to be delivered –
until nine. However tired we are, we are not allowed to go home. We may
get an extra 200 rupiah (10 US cents) . . . We go on foot to the factory from
where we live. Inside it is very hot. The building has a metal roof, and there
is not much space for all the workers. It is very cramped. There are over 200
people working there, mostly women, but there is only one toilet for the
whole factory . . . when we come home from work, we have no energy left to
do anything but eat and sleep.'   (Seabrook, 1996, 90–1)

Home is a single room, 2 metres by 3, costing $16 a month; it costs nearly
10 cents to get two cans of water and at least a $1.50 a day to eat.

In *Capital* Marx recounts the story of the milliner, Mary Anne Walkley,
twenty years of age, who often worked 30 hours without a break (though
revived by occasional supplies of sherry, port, and coffee) until, after a
particularly hard spell necessitated by preparing 'magnificent dresses for
the noble ladies invited to the ball in honour of the newly imported
Princess of Wales,' died, according to the doctor's testimony, 'from long
hours of work in an over-crowded work-room, and a too small and badly
ventilated bedroom' (1976 edition, 364). Compare that with a contemporary account of conditions of labor in Nike plants in Vietnam:

[Mr Nguyen] found that the treatment of workers by the factory managers
in Vietnam (usually Korean or Taiwanese nationals) is a 'constant source of
humiliation,' that verbal abuse and sexual harassment occur frequently, and
that 'corporal punishment is often used.' He found that extreme amounts of
forced overtime are imposed on Vietnamese workers. 'It is a common
occurrence,' Mr Nguyen wrote in his report, 'to have several workers faint
from exhaustion, heat and poor nutrition during their shifts. We were told
that several workers even coughed up blood before fainting.' Rather than
crack down on the abusive conditions in the factories, Nike has resorted to
an elaborate international public relations campaign to give the appearance
that it cares about its workers. But no amount of public relations will change
the fact that a full-time worker who makes $1.60 a day is likely to spend a
fair amount of time hungry if three very simple meals cost $2.10.
                                                              (Herbert, 1997)

The material conditions that sparked the moral outrage that suffuses the
Manifesto have not gone away. They are embodied in everything from
Nike shoes, Disney products, GAP clothing to Liz Claiborne products.
And, as in the nineteenth century, part of the response has been reformist
middle-class outrage backed by the power of working-class movements to

regulate conditions of labor worldwide (Moody, 1997). Campaigns against 'sweatshop labor' worldwide and for a code of 'fair labor practices' perhaps certified by a 'fair labor label' on the products we buy, as well as the specific campaigns against Nike and other major corporations, are a case in point (Ross, 1997; Goodman, 1996; Greenhouse, 1997a; 1997b).

The setting for the *Manifesto* has not, then, radically changed at its basis. The global proletariat is far larger than ever and the imperative for workers of the world to unite is greater than ever. But the barriers to that unity are far more formidable than they were in the already complicated European context of 1848. The workforce is now far more geographically dispersed, culturally heterogeneous, ethnically and religiously diverse, racially stratified, and linguistically fragmented. The effect is to radically differentiate both the modes of resistance to capitalism and the definitions of alternatives. And while it is true that means of communication and opportunities for translation have greatly improved, this has little meaning for the billion or so workers living on less than a dollar a day possessed of different cultural histories, literatures, and understandings (compared to international financiers and transnationals who use the new forms of telecommunications all the time).

Differentials (both geographical and social) in wages and social provision within the global working class are likewise greater than they have ever been. The political and economic gap between the most affluent workers, in say Germany and the United States, and the poorest wage workers, in Indonesia and Mali, is far greater than between the so-called aristocracy of European labor and their unskilled counterparts in the nineteenth century. This means that a certain segment of the working class (mostly but not exclusively in the advanced capitalist countries and often possessing by far the most powerful political voice) has a great deal to lose besides its chains.

While women were always an important component of the workforce in the early years of capitalist development, their participation has now become much more general at the same time as it has become concentrated in certain occupational categories (usually dubbed 'unskilled' – see Wright, 1996). This poses acute questions of gender in working-class politics that have too often been pushed under the rug in the past. On this point, the *Manifesto* is less than convincing. To be sure, there are many astute observations on the forces set fair to destroy the family as an institution and to transform it into a commodified property relation. And Marx and Engels certainly declared their aim was 'to do away with the status of women as mere instruments of production.' They also sought, contrary to bourgeois fears, to abolish what they call 'the community of women' deriving from private property (a system of 'prostitution, both

public and private') (70–1). But this still leaves little room for broader
forms of gender struggle within the socialist program. A strongly feminized proletarian movement (not an impossibility in our times) might turn
out to be a different agent of political transformation to that led almost
exclusively by men.

Ecological variations and their associated impacts (resource wars,
environmental injustice, differential effects of environmental degradation)
have become far more salient in the quest for an adequate quality of life as
well as for rudimentary health care. In this regard, too, there is no level
playing field upon which class struggle can be evenly played out because
the relation to nature is itself a cultural determination that can have
implications for how any alternative to capitalism can be constructed.
While such cross-cultural conditions provide a basis for a radical critique
of the purely utilitarian and instrumental attitudes towards the natural
world embedded in capitalist accumulation practices, they also pose a
challenge to the socialist movement to define a more ecologically sensitive
politics than has often been proposed in the past. How to configure the
environmental with the economic, the political with the cultural, becomes
much harder at the global level, where presumptions of homogeneity of
values and aspirations across the earth simply do not hold.

Global populations have also been on the move. The flood of migratory
movements seems impossible to stop. State boundaries are less porous for
people and for labor than they are for capital, but they are still porous
enough. Immigration is a significant issue worldwide (including within the
labor movement itself). Organizing labor in the face of the considerable
ethnic, racial, religious, and cultural diversity generated out of migratory
movements poses particular problems that the socialist movement has
never found easy to address let alone solve. Europe, for example, now has
to face all of those difficulties that stem from ethnic and racial diversity
that have been wrestled with for many years (and often proved so divisive
to working-class unities) in the United States.

Urbanization has also accelerated to create a major ecological, political,
economic, and social revolution in the spatial organization of the world's
population. The proportion of an increasing global population living in
cities has doubled in thirty years, making for massive spatial concentrations of population on a scale hitherto regarded as inconceivable. It has
proven far easier to organize class struggle in, say, the small-scale mining
villages of the South Wales Coalfield, or even in relatively homogeneous
industrial cities like nineteenth-century Manchester (with a population of
less than a million, albeit problematically divided between English and
Irish laborers), than organizing class struggle (or even developing the
institutions of a representative democracy) in contemporary Sao Paulo,

Cairo, Lagos, Los Angeles, Shanghai, Bombay, etc. with their teeming, sprawling, and often disjointed populations reaching close to or over the twenty-million mark. The geographical scale at which places are now defined (see Plate 3.1) is so different as to make the whole prospect of politics quite different.

The socialist movement has to come to terms with these extraordinary geographical transformations and develop tactics to deal with them. This does not dilute the importance of the final rallying cry of the *Manifesto* to unite. The conditions that we now face make that call more imperative than ever. But we cannot make either our history or our geography under historical-geographical conditions of our own choosing. The geographical reading of the *Manifesto* laid out in Chapter 2 emphasizes the non-neutrality of spatial structures and powers in the intricate spatial dynamics of class struggle. It reveals how the bourgeoisie acquired its powers *vis-à-vis* all preceding modes of production by mobilizing command over

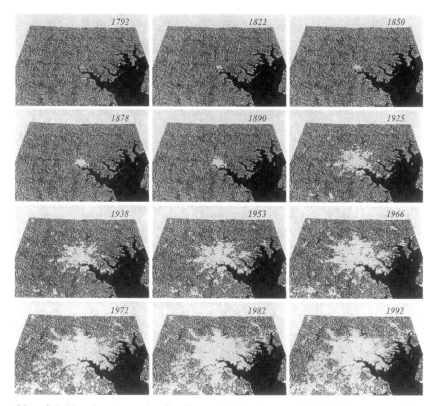

Plate 3.1 The changing scale of a place called 'Baltimore,' 1792–1992. This perspective view of urban growth in Baltimore, Maryland, over a 200-year period illustrates the problem of changing scale in urban organization. It poses the question: in what ways does it make sense to consider 'Baltimore' a consistent political, economic or ecological unit of analysis over time?

space as a productive force peculiar to itself. It shows how the bourgeoisie has continuously enhanced and protected its power by that same mechanism. It therefore follows that until the working class movement learns how to confront that bourgeois power to command and produce space, to shape a new geography of production and social relations, it will always play from a position of weakness rather than of strength. Likewise, until that movement comes to terms with the geographical as well as the historical conditions and diversities of its own existence, it will be unable to define, articulate, and struggle for a realistic socialist alternative to capitalist domination.

The *Manifesto* clearly states that the task of Communists is to 'point out and bring to the front the common interests of the entire proletariat, independently of all nationality' as well as to represent, without setting up sectarian interests of their own (an injunction all too often ignored in the past), 'the interests of the movement as a whole.'

> [Communists should aspire to be] on the one hand, practically, the most advanced and resolute sections of the working class parties of every country ... on the other hand, theoretically, they have over the mass of the proletariat the advantage of clearly understanding the line of march, the conditions, and the ultimate general results of the proletarian movement.
>
> (62)

There is, in this, a danger of a certain arrogance, a presumption that the laws of history (and of geography?) are known to us and only us and that we and only we understand the proper lines of political action. The science of Marxism, powerful and insightful though it undoubtedly is, cannot claim to omniscience nor is it lacking in its uncertainties. There is, furthermore, the problem of whose imagination is to prevail in the construction of any alternative. The socialist utopians that preceded Marx and Engels produced 'fantastic pictures of future society, painted at a time when the proletariat [was] still in an undeveloped state and has but a fantastic conception of its own position.' These pictures correspond to 'the first instinctive yearnings of that class for a general reconstruction of society.' Furthermore, the socialist utopians produced a repertoire of critical principles and 'valuable materials for the enlightenment of the working class' (91). The problem, Marx and Engels aver, is that the development of the class struggle itself renders this utopianism redundant, even turns it into a negative force as sects form around charismatic leaders who seek their own 'spatial fix' to social problems by founding isolated communities or colonies overseas. Marx therefore pleaded, as we have seen, with the Icarians, as 'good communists,' not to abandon the class struggle in Europe and to flee to their utopia.

But, in a time when the class struggle has receded as a unified force in the advanced capitalist world (though still present in a thousand and one fragmented forms), is this not also a time when the painting of fantastic pictures of a future society has some role to play? We desperately need a revitalized socialist avant-garde, an international political movement capable of bringing together in an appropriate way the multitudinous discontents that derive from the naked exercise of bourgeois power in pursuit of a utopian neoliberalism. This does not mean an old-style avant-garde party that imposes a singular goal and arrogates to itself such clarity of vision as to exclude all other voices. But it does mean the creation of organizations, institutions, doctrines, programs, formalized structures, and the like, that work to some common purpose. Such political activities must be firmly grounded in the concrete historical and geographical conditions under which human action unfolds. Between the traditional avant-gardism of communist parties (the specter of the *Manifesto* given form by Lenin) and the idealized avant-gardism dominant in the academy – the specter of someone like Derrida (1994) – there lies a terrain of political organization and struggle that desperately cries out for cultivation. That terrain is not, fortunately, empty of possibilities.

Some clues as to strategies with which to confront this problem are embedded in the *Manifesto* itself. Properly embellished, such insights can take us onto richer terrains of struggle. It is important to accept, for example, that the beginning point of class struggle lies with the particularity of the laboring body, with figures like Mary Anne Walkley and Hira and Mira and the billions of others whose daily existence is shaped through an often traumatic and conflictual relation to the dynamics of capital accumulation. The laboring body is, therefore, a site of resistance that achieves a political dimension through the political capacity of individuals to act as moral agents (see Chapter 7). To treat of matters this way is not to revert to some rampant individualism but to insist, as the *Manifesto* does, that the universality of class struggle originates with the particularity of the person and that class politics must translate back to that person in meaningful ways. The alienation of the individual is, therefore, an important beginning point for politics and it is that alienation that must be overcome.

But, and this is of course the crucial message of the *Manifesto*, that alienation cannot be addressed except through collective struggle. That means building a movement that reaches out across space and time in such a way as to confront the universal and transnational qualities of capital accumulation. Ways have to be found to connect the microspace of the body with the macrospace of what is now called 'globalization.' The *Manifesto* suggests this can be done by linking the personal to the local to the regional, the national, and ultimately the international. A hierarchy of spatial scales exists at which class politics must be constructed. But the 'theory of the production of geographical scale,' as Smith (1992) observes, 'is grossly underdeveloped' and we have yet to learn, particularly with respect to global working class formation and body politics, how to 'arbitrate and translate' between the different spatial scales. This is an acute problem that must be confronted and resolved if working-class politics is to be revived. I give just three examples.

The traditional beginning point for class struggle has been a particular space – the factory – and it is from there that class organization has been built up through union movements, political parties, and the like. But what happens when factories disappear or become so mobile as to make permanent organizing difficult if not impossible? And what happens when much of the workforce becomes temporary or casualized? Under such conditions labor organizing in the traditional manner loses its geographical basis and its powers are correspondingly diminished. Alternative models of organizing must then be constructed. In Baltimore, for example, there is a city-wide movement for a living wage (see Chapter 7) that relies upon alliances of institutions of community (particularly the churches), activist organizations, and student groups, as well as whatever union support can be procured, to achieve its objectives. A movement is created across the whole metropolitan space that operates outside of traditional labor organizing models but in a way that addresses new conditions (particularly, as Herod [1998] stresses, the geographical scale at which labor organization is required). This is a version of what Moody (1997) calls 'social unionism' at work within the politics of place.

Consider a second example. Governmentality for contemporary capitalism has entailed the construction of important supra-national authorities such as NAFTA and the European Union. Unquestionably, such constructions – the Maastricht Agreement being the paradigmatic case – are pro-capitalist. How should the left respond? The divisions here are important to analyze (in Europe the debate within the left is intense), but too frequently the response is an overly-simplistic argument that runs along the following lines: 'because NAFTA and Maastricht are pro-capitalist we fight them by defending the nation state against supra-national governance.' The argument here outlined suggests an entirely different response. The left must learn to fight capital at *both* spatial scales simultaneously. But, in so doing, it must also learn to coordinate potentially contradictory politics within itself at the different spatial scales for it is often the case in hierarchical spatial systems (and ecological problems frequently pose this dilemma) that what makes good political sense at one scale does not make such good politics at another (the rationalization of, say, automobile production in Europe may mean plant closures in Oxford or Turin). Withdrawing to the nation state as the exclusive strategic site of class organization and struggle is to court failure (as well as to flirt with nationalism and all that that entails). This does not mean the nation state has become irrelevant – indeed it has become more relevant than ever. But the choice of spatial scale is not 'either/or' but 'both/and' even though the latter entails confronting serious contradictions. This means that the union movement in the United States ought to put just as much effort into cross-border organizing (particularly with respect to Mexico) as it puts into fighting NAFTA and that the European union movement must pay as much attention to procuring power and influence in Brussels and Strasbourg as each member of the movement does in its own respective national capital.

Moving to the international level poses similar dilemmas and problems. It is interesting to note that the internationalism of labor struggle, while it hovers as an obvious and latent necessity over much of the labor movement, faces serious difficulties organizationally. I again in part attribute this to a failure to confront the dilemmas of integrating struggles at different spatial scales. Examples exist of such integrations in other realms. Movements around human rights, the environment, and the condition of women illustrate the possible ways in which politics can get constructed (as well as some of the pitfalls to such politics) to bridge the micro-scale of the body and the personal on the one hand and the macro-scale of the global and the political-economic on the other. Nothing analogous to the Rio Conference on the environment or the Beijing Conference on women has occurred to confront global conditions of labor. We have scarcely begun to think of concepts such as 'global working class formation' or even to analyze what that might mean. Much of the defense of human dignity in the face of the degradation and violence of labor worldwide has been articulated through the churches and human rights organizations rather than through labor organization directly (the churches' ability to work at different spatial scales provides a number of models for political organization from which the socialist movement could well draw some important lessons). As in the case of local-level struggles, alliances between labor organizations and many other institutions in civil society appear now to be crucial to the articulation of socialist politics at the international scale. Many of the campaigns orchestrated in the United States, for example, against global sweatshops in general or particular versions (such as Disney operations in Haiti and Nike worldwide) are organized through such alliances (see Ross, 1997). The argument here is not that nothing is being done or that institutions do not exist (Moody, 1997, provides several examples of formative institutions of international labor organization at work). But the reconstruction of some sort of socialist internationalism after 1989 has not been an easy matter even if the collapse of the Wall opened up new opportunities to explore that internationalism free of the need to defend the rump of the Bolshevik Revolution against the predatory politics of capitalist powers.

How to build a political movement at a variety of spatial scales as an answer to the geographical and geopolitical strategies of capital is a problem that in outline at least the *Manifesto* clearly articulates. How to do it for our times is an imperative issue for us to resolve. We cannot set about that task without recognizing the geographical complexities that confront us. We need to find ways to construct a dialectics of politics that moves freely from the micro- to macro-scales and back again. The clarifications that a study of the *Manifesto*'s geography offer provide a marvelous opportunity to wrestle with that task in such a way as to reignite the flame of socialism from Jakarta to Los Angeles, from Shanghai to New York City, from Porto Alegre to Liverpool, from Cairo to Warsaw, from Beijing to Turin.

There is no magic answer. But there is at least a strategic way of thinking that can illuminate the way. This is what the 1848 *Manifesto* still provides. Above all, the political passions that suffuse it are an incredible inspiration. It still reads as an extraordinary document, full of insights, rich in meanings and bursting with political possibilities. Marx and Engels created a text, produced a captivating rhetorical form, that has been read, appreciated and absorbed (though often in undefinable ways) by millions of people around the world for more than 150 years. How we read it now and appropriate its meaning today is a crucial question for progressive politics, always remembering, as the *Manifesto* itself reminds us, that the practical applications of its principles will always depend 'on the conditions for the time being existing' (8).

CHAPTER 4

Contemporary globalization

Over the last twenty years or so, 'globalization' has become a key word for organizing our thoughts as to how the world works. How and why it came to play such a role is in itself an interesting tale. I want here, however, to focus on the theoretical and political implications of the rise of such a concept. I begin with two general sets of questions in order to highlight what appear to be important political changes in western discourses (though not necessarily in realities).

1. Why is it that the word 'globalization' has recently entered into our discourses in the way it has? Who put it there and why and by means of what political project? And what significance attaches to the fact that even among many 'progressives' and 'leftists' in the advanced capitalist world, much more politically loaded words like 'imperialism,' 'colonialism,' and 'neocolonialism' have increasingly taken a back seat to 'globalization' as a way to organize thoughts and to chart political possibilities?

2. How has the conception of globalization been used politically? Has adoption of the term signalled a confession of powerlessness on the part of national, regional, and local working-class or other anti-capitalist movements? Has belief in the term operated as a powerful deterrent to localized and even national political action? Has the form of solidarity hitherto represented by the nation state become 'hollowed out' as some now claim? Are all oppositional movements to capitalism within nation states and localities such insignificant cogs in the vast infernal global machine of the international market place that there is no room for political maneuver anywhere?

Viewed from these perspectives, the term globalization and all its associated baggage is heavily laden with political implications that bode ill for most traditional forms of leftist or socialist politics. But before we reject or abandon it entirely, it is useful to take a good hard look at what it incorporates and what we can learn, theoretically and politically, from the brief history of its use.

1 Globalization as process

Globalization can be viewed as a process, as a condition, or as a specific kind of political project. These different approaches to it are not, as I shall hope to show, mutually exclusive. But I propose to begin by considering it as a process. To view it this way does not presume that the process is constant, nor does it preclude saying that the process has, for example, entered into a radically new stage or worked itself out to a particular or even 'final' condition. Nor does it 'naturalize' globalization as if it has come about without distinctive agents working to promote it. But taking up the process-based angle makes us concentrate in the first instance on *how* globalization has occurred and is occurring.

What we then see is that something akin to 'globalization' has a long presence within the history of capitalism. Certainly from 1492 onwards, and even before, the internationalization of trade and commerce was well under way. Capitalism cannot do without its 'spatial fixes' (cf. Chapter 2). Time and time again it has turned to geographical reorganization (both expansion and intensification) as a partial solution to its crises and impasses. Capitalism thereby builds and rebuilds a geography in its own image. It constructs a distinctive geographical landscape, a produced space of transport and communications, of infrastructures and territorial organizations, that facilitates capital accumulation during one phase of its history only to have to be torn down and reconfigured to make way for further accumulation at a later stage. If, therefore, the word 'globalization' signifies anything about our recent historical geography, it is most likely to be a new phase of exactly this same underlying process of the capitalist production of space.

It is not my intention to review the vast literature that deals with the spatial and geographical aspects of capitalist development and class struggle (even if such a task were feasible). But I do think it important to recognize a series of tensions and often uncomfortable compromises as to how to understand, theoretically and politically, the geographical dynamics of capital accumulation and class struggle. When, for example, Lenin and Luxemburg clashed on the national question, as the vast controversy on the possibility of socialism within one country (or even within one city) unfolded, as the Second International compromised with nationalism in the First World War, and as the Comintern subsequently swayed back and forth on how to interpret its own internationalism, so the socialist/ communist movement never managed to evolve, politically or theoretically, a proper or satisfactory understanding of how the production of space was fundamental and integral to the dynamics of capital accumulation and the geopolitics of class struggle.

A study of the *Manifesto* (Chapter 2) indicates one key source of the dilemma. For while it is clear that the bourgeoisie's quest for class domination was (and is) a very geographical affair, the almost immediate reversion in the text to a purely temporal and diachronic account is striking. It is hard, it seems, to be dialectical about space, leaving many Marxists in practice to follow Feuerbach in thinking that time is 'the privileged category of the dialectician, because it excludes and subordinates where space tolerates and coordinates.' (Ross, 1988, 8). Even the term 'historical materialism,' I note, erases the significance of geography and if I have struggled these last few years to try to implant the idea of 'historical-geographical materialism' it is because the shift in that terminology prepares us to look more flexibly and, I hope, more cogently at the class significance of processes like globalization and uneven geographical development (Harvey, 1996). We need far better ways to understand if not resolve politically the underlying tension between what often degenerates into either a temporal teleology of class triumphalism (now largely represented by the triumphalism of the bourgeoisie declaring the victorious end of history) or a seemingly incoherent and uncontrollable geographical fragmentation of class and other forms of social struggle in every nook and cranny of the world.

Since Marx, for example, a variety of accounts has been offered of how capitalism has structured its geography (such as Lenin's theory of imperialism, Luxemburg's positioning of imperialism as the savior of capitalist accumulation, Mao's depiction of primary and secondary contradictions in class struggle). These have subsequently been supplemented by more synthetic accounts of accumulation on a world scale (Amin, 1974), the production of a capitalist world system (Wallerstein, 1974; Arrighi, 1994), the development of underdevelopment (Frank, 1969, and Rodney, 1981), unequal exchange (Emmanuel, 1972) and dependency theory (Cardoso and Faletto, 1979). As Marxist ideas and political practices have spread throughout the globe (in a parallel process of globalization of class struggle) so innumerable local/national accounts of resistance to the invasions, disruptions, and imperialist designs of capitalism have been generated.

We have then to recognize the geographical dimension and grounding for class struggle. As Raymond Williams (1989, 242) suggests, politics is always embedded in 'ways of life' and 'structures of feeling' peculiar to places and communities. The universalism to which socialism aspires has, therefore, to be built by negotiation between different place-specific demands, concerns, and aspirations. It has to deal with what he called 'militant particularism.' By this he meant:

> The unique and extraordinary character of working class self-
> organization ... to connect particular struggles to a general struggle in one
> quite special way. It has set out, as a movement, to make real what is at first
> sight the extraordinary claim that the defence and advancement of certain
> particular interests, *properly brought together*, are in fact the general interest.
>   (my emphases)

Even temporal accounts of class struggle turn out to be territorially
bounded. But there has been little concern to justify the geographical
divisions upon which such accounts are based. We then have innumerable
accounts of the making of the English, Welsh, French, German, Italian,
Catalan, South African, South Korean, etc. working classes as if these are
natural geographical entities. Attention focuses on class development
within some circumscribed space which, when scrutinized more closely,
turns out to be a space within an international space of flows of capital,
labor, information, etc., in turn comprised of innumerable smaller spaces
each with its own distinctive regional or even local characteristics. When
we look closely at the action described in Edward Thompson's magisterial
account of *The Making of the English Working Class*, for example, it turns
out to be a series of highly localized events often loosely conjoined in space.
John Foster may have rendered the differences somewhat too mechanical
in his own account of *Class Struggle in the Industrial Revolution* but it is, I
think, undeniable that class structure, class consciousness, and class
politics in Oldham, Northampton, and South Shields (read Colmar, Lille,
St. Etienne, or Minneapolis, Mobile, and Lowell) were quite differently
constructed and worked out, making geographical differences within the
nation state rather more important than most would want to concede.

This mode of thinking uncritically about supposedly 'natural' geographical entitities is now perpetuated in neo-Marxist accounts of capital
(particularly those inspired by 'regulation theory') that make it seem as if
there are distinctive German, British, Japanese, American, Swedish,
Singaporean, Brazilian, etc. versions of capitalism (sometimes broken
down into more regionalized orderings, such as North versus South in
Italy, Brazil, Britain ...). These distinctive capitalisms are then construed
as entities in competition with each other within a global space economy.
This conception is by no means confined to the left. It is standard
procedure across a wide variety of political positions to compare different
national capitalisms (Japanese, Nordic, German, etc.) as if they are meaningful entities.

My argument here is not that these national or cultural distinctions are
wrong, but that they are so easily presumed to exist without the assem-
blage of any evidence or argument for them whatsoever. They are held, as
it were, to be self-evident, when a little probing shows that they are either

far more complicated than is assumed or so fuzzy and porous as to be highly problematic. The concept of 'place' that Williams invokes turns out to be more complicated than he imagined. The result is a clear line of tension within most accounts of recent political-economic changes. On the one hand, we have spaceless and geographically undifferentiated accounts (mainly theoretical these days though polemical and political versions also abound mainly in right-wing and conservative incarnations) that understand capitalist development as a purely temporal process moving inexorably towards some given destination. In the traditional left-wing version class struggle powers the historical movement towards socialism/communism as an (inevitable) outcome. On the other hand, we have geographical accounts in which class alliances (and this often includes a working class characterized by what Lenin condemned as a limiting trade-union consciousness) form within places to exploit class alliances in other places (with, perhaps, a *comprador* bourgeoisie as agent). In this case imperialism (or, conversely, struggles for national liberation or local autonomy) hold the key to the future. The two accounts are somehow supposed to be consistent with each other. In fact the theoretical justification for viewing the exploitation of one class by another as homologous with the exploitation of a class alliance in one place by another has never been strong. And the assumption that struggles to liberate spaces (e.g., struggles for national liberation) are progressive in the class-struggle sense (for either a nascent bourgeoisie or a working class) cannot stand up to strong scrutiny. There are numerous examples of class and national liberation struggles confounding each other. How, then, can we unconfound this problem?

One of the things that the rise of the term 'globalization' to pre-eminence signals is a profound geographical reorganization of capitalism, making many of the presumptions as regards the 'natural' geographical units within which capitalism's historical trajectory develops less and less meaningful (if they ever were). We are therefore faced with a historic opportunity to seize the nettle of capitalism's geography, to see the production of space as a constitutive moment within (as opposed to something derivatively constructed by) the dynamics of capital accumulation and class struggle. This provides an opportunity to emancipate ourselves from imprisonment within a hidden spatiality that has had the opaque power to dominate (and sometimes to confuse) the logic of both our thinking and our politics. It also permits us to understand better exactly how class and inter-place struggles so often cut across each other and how capitalism can frequently contain class struggle through a geographical divide and rule of that struggle. We are then in a better position to understand the spatio-temporal contradictions inherent in capitalism and, through that understanding, better speculate on how to exploit the weakest link and so explode the worst horrors of capitalism's penchant for violent though 'creative' destruction.

How, then, can we dance to this agenda, both theoretically and politically?

There are, of course, innumerable signs of a willingness to take on the theoretical implications of changing spatialities and re-territorializations. A prime virtue of Deleuze and Guattari's *Anti-Oedipus*, for example, was to point out that the territorialization and reterritorialization of capitalism is an on-going process. But here, as in many other accounts, the virtue of a re-spatialization of social thought has been bought at the cost of partial and sometimes radical breaks with Marxist formulations (both theoretical and political). In my own work, I have sought to show that there are ways to integrate spatialities into Marxist theory and practice without necessarily disrupting central propositions, though in the course of such an integration all sorts of modifications to both theory and practice do arise. So let me summarize some of the main features of this argument.

I begin with the simplest propositions I can find. Certain tensions are embedded within any materialist account of the circulation process of capital and the organization of labor processes to extract surplus value. These periodically and inescapably erupt as powerful moments of historical-geographical contradiction.

First, capitalism is always under the impulsion to accelerate turnover time, to speed up the circulation of capital and consequently to revolutionize the time horizons of development. But it can do so only through long-term investments (in, e.g., the built environment as well as in elaborate and stable infrastructures for production, consumption, exchange, communication, and the like). A major strategem of crisis avoidance, furthermore, lies in absorbing overaccumulated capital in long-term projects (e.g., the famous 'public works' launched by the state in times of depression) and this slows down the turnover time of capital. There is, consequently, an extraordinary array of contradictions that collect around the issue of the time-horizon within which different capitals function. Historically, and now is no exception, this tension has primarily been registered through the contradictions between money and finance capital (where turnover is now almost instantaneous), on the one hand, and merchant, manufacturing, agrarian, information, construction, service, and state capitals on the other. But contradictions can be found within factions (between currency and bond markets, for example, or between landlords, land developers, and speculators). All sorts of mechanisms exist for coordinating between capital dynamics working to different temporal rhythms. But uneven development of turnover times and temporalities, of the sort produced by the recent implosion of time-horizons in a powerful financial sector, can create an unwelcome temporal compression that is deeply stressful to other factions of capital, including, of course, that embodied in the capitalist state. The time-horizon set by Wall Street simply cannot accommodate to the temporalities of social and ecological reproduction systems in a responsive way. And it goes without saying that the rapid turnover time set in financial markets is even more stressful for workers (their job security, their skills, etc.) and for the lifeworld of socio-ecological reproduction. This stress-point has been central to the political economy of advanced capitalism these last twenty years.

Second, capitalism is under the impulsion to eliminate all spatial barriers, to 'annihilate space through time' as Marx puts it, but it can do so only through the production of a fixed space. Capitalism thereby produces a geographical landscape (of space relations, of territorial organization, and of systems of places linked in a 'global' division of labor and of functions) appropriate to its own dynamic of accumulation at a particular moment of its history, only to have to destroy and rebuild that geographical landscape to accommodate accumulation at a later date. There are a number of distinct aspects to this process.

1. Reductions in the cost and time of movement over space have been a continuing focus of technological innovation. Turnpikes, canals, railroads, electric power, the automobile, and air and jet transport have progressively liberated the movements of commodities and people from the constraints of the friction of distance. Parallel innovations in the postal system, the telegraph, the radio, telecommunications, and the worldwide web have now pushed the cost of transfer of information (though not of the infrastructures and terminals) close to zero.

2. The building of fixed physical infrastructures to facilitate this movement as well as to support the activities of production, exchange, distribution, and consumption exercises a quite different force upon the geographical landscape. More and more capital is embedded in space as landed capital, as capital fixed in the land, creating a 'second nature' and a geographically-organized resource structure that more and more inhibits the trajectory of capitalist development. The idea of somehow dismantling the urban infrastructures of Tokyo-Yokohama or New York City overnight and starting all over again is simply ludicrous. The effect is to make the geographical landscape of capitalism more and more sclerotic with time, thus creating a major contradiction with the increasing liberty of movement. That tendency is made even more emphatic to the degree that the institutions of place become strongly articulated and loyalties to places (and their specific qualities) become a significant factor in political action.

3. The third element is the construction of territorial organization, primarily (though not solely) state powers to regulate money, law, and
   politics and to monopolize the means of coercion and violence according
   to a sovereign territorial (and sometimes extra-territorial) will. There
   are, of course, innumerable Marxist theories of the state, many of which
   engage in an unhealthy degree of abstraction from history and geography, making it seem as if states like Gabon and Liberia are on a par with
   the United States or Germany, and failing to recognize that most of the
   state boundaries in the world were drawn between 1870 and 1925. In
   Europe there were more than 500 political entities in 1500 reduced to 23
   by 1920. In recent years the number has increased to 50 or more with
   several further secessions threatened. Most states in the world became
   independent only after 1945 and many of them have been in search of a
   nation ever since (but then this was as historically true of France and
   Mexico as it has recently been of Nigeria or Rwanda). So while it is true
   that the Treaty of Westphalia established for the first time in 1648 the
   principle that independent sovereign states, each recognizing their
   autonomy and territorial integrity, should co-exist in the capitalist
   world, the process of territorializing the world according to that
   principle took several centuries to complete (accompanied by a good
   deal of violence). And the processes that gave rise to that system can just
   as easily dissolve it, as some commentators are now arguing is indeed
   happening as supra-national organizations (such as the European Union) and regional autonomy movements within nation states do their
   work. In short, we have to understand the processes of state formation
   and dissolution in terms of the unstable processes of globalization/
   territorialization. We then see a process of territorialization, de-terri-
   torialization, and re-territorialization continuously at work throughout
   the historical geography of capitalism (this was one of the fundamental
   points that Deleuze and Guattari reinforced in *Anti-Oedipus*).

Armed with these concepts we can, I think, better understand the process
of globalization as a process of production of uneven temporal and
geographical development. And, as I shall hope to show, that shift of
language can have some healthy political consequences, liberating us from
the more oppressive and confining language of an omnipotent and
homogenizing process of globalization.

2 Recent shifts in the dynamics of globalization

Bearing that in mind, let me come back to what the term 'globalization'
might signify and why it has taken on a new allure and thereby become so
important in recent times. Four shifts stand out:

1. Financial deregulation began in the USA in the early 1970s as a forced response to the stagflation then occurring internally and to the breakdown of the Bretton Woods System of international trade and exchange (largely because of the uncontrolled growth of the Eurodollar market). I think it important to recognize that the wave of financial deregulation was less a deliberate strategy thought out by capital than a concession to reality (even if certain segments of capital stood to benefit far more than others). Bretton Woods had been a global system too, so what happened here was a shift from a global system that was hierarchically organized and largely controlled by the United States to another global system that was more decentralized, coordinated through the market, and made the financial conditions of capitalism far more volatile. The rhetoric that accompanied this shift (which occurred in a series of steps from 1968 onwards but most particularly from 1979 to 1985) was implicated in the promotion of the term 'globalization' as a virtue. In my more cynical moments I find myself thinking that it was the financial press that conned us all (me included) into believing in 'globalization' as something new when it was nothing more than a promotional gimmick to make the best of a necessary adjustment in the system of international finance. Coincidentally the financial press has for some time now been stressing the importance of the regionalization in financial markets, the Japanese co-prosperity sphere, NAFTA, and the European Union being the obvious power blocs – sometimes referred to as 'the Triad'. Some even among the boosters of globalization have been warning that the 'backlash' against globalization, mainly in the form of populist nationalisms, is to be taken seriously, and that globalization is in danger of becoming synonymous with 'a brakeless train wreaking havoc' (Friedman, 1996).

2. The waves of profound technological change and product innovation and improvement that have washed over the world since the mid-1960s provide a major focus for any enquiry into recent transformations in the world economy. There have, of course, been many similar phases of technological innovation before in the long history of capitalism. Innovations tend to bunch together (for a variety of often synergistic reasons). We have certainly been living through such a concentrated period of change in recent times. But what may be more special now is the pace and rate of technology transfer and immitation across and throughout different zones of the world economy. Some of this has to do with a global arms trade but the existence of educated and scientifically-trained elites capable of adapting and absorbing technological knowledge and know-how from anywhere and everywhere also has something to do with the rapidity with which new technologies and

products diffuse throughout the world (the problem of nuclear profusion is indicative here). For this reason many now consider galloping technological innovation and transfer as the most singular and seemingly unstoppable force promoting globalization.

3. The media and communications system and, above all, the so-called 'information revolution' brought some significant changes to the organization of production and consumption as well as to the definition of entirely new wants and needs. The ultimate 'de-materialization of space' in the communications field had its origins in the military apparatus, but was immediately seized upon by financial institutions and multinational capital as a means to coordinate their activities instantaneously over space. The effect has been to form a so-called de-materialized 'cyberspace' in which certain kinds of important transactions (primarily financial and speculative) could occur. But then we also came to watch revolutions and wars live on television. The space and time of media and communications imploded in a world where the monopolization of media power has become more and more of a problem (in spite of proclamations of libertarian democratization via the internet).

The idea of an 'information revolution' is powerfully present these days and is often viewed as the dawning of a new era of globalization within which the information society reigns supreme (see e.g. Castells, 1996). It is easy to make too much of this. The newness of it all impresses, but then the newness of the railroad and the telegraph, the automobile, the radio, and the telephone in their day impressed equally. These earlier examples are instructive since each in their own way did change the way the world works, the ways in which production and consumption could be organized, politics conducted, and the ways in which social relations between people could become converted on an ever widening scale into social relations between things. And it is clear that the relations between working and living, within the workplace, in cultural forms, are indeed changing rapidly in response to informational technology. Interestingly, this is a key component in the right-wing political agenda in the United States. The new technology, said Newt Gingrich (advised by Alvin Toffler whose right-wing utopianism rests entirely on the idea of a 'third-wave' information revolution), is inherently emancipatory. But in order to liberate this emancipatory force from its political chains it is essential to pursue a political revolution to dismantle all of the institutions of 'second wave' industrial society – government regulation, the welfare state, collective institutions of wage bargaining, and the like. That this is a vulgar version of the Marxist argument that changes in productive forces drive social relations and history should not be lost upon us. Nor should we ignore the strong teleological tone to this right-wing rhetoric (perhaps best captured in Margaret Thatcher's famous declaration that 'there is no alternative').

4. The cost and time of moving commodities and people also tumbled in another of those shifts that have periodically occurred within the history of capitalism. This liberated all sorts of activities from former spatial constraints, permitting rapid adjustments in locations of production, consumption, populations, and the like. When the history of the globalization process comes to be written, this simple shift in the cost of overcoming space may be seen as far more significant than the so-called information revolution *per se* (though both are part and parcel of each other in practice).

It is, perhaps, invidious to take these elements separately because in the end it is probably the synergistic interactions between them which is of the greatest import. Financial deregulation could not have occurred, for example, without the information revolution, and technology transfer (which also relied heavily on the information revolution) would have been meaningless without a much greater ease of movement of commodities and people around the world.

3 Consequences and contradictions

These four shifts in the globalization process were accompanied by a number of other important features, perhaps best thought of as derivative from the primary forces at work.

1. Production and organizational forms (particularly of multinational capital, though many small entrepreneurs also seized new opportunities) changed, making abundant use of the reduced costs of commodity and information movement. Offshore production that began in the 1960s suddenly became much more general (it has now spread with a vengeance even to Japan as production moves to China or other areas of South-East Asia). The geographical dispersal and fragmentation of production systems, divisions of labor, and specializations of tasks ensued, albeit often in the midst of an increasing centralization of corporate power through mergers, takeovers, or joint production agreements that transcended national boundaries. Corporations, though many of them still retain a strong base in a home country (few are genuinely transnational), have more power to command space, making individual places much more vulnerable to their whims. The global television set, the global car, became an everyday aspect of political-economic life. The closing down of production in one place and the

opening up of production somewhere else became a familiar story –
some large-scale production operations have moved four or five times
in the last twenty years.

2. The world wage labor force more than doubled in less than twenty
    years (see Chapter 3). This occurred in part through rapid population
    growth but also through bringing more and more of the world's
    population (particularly women) into the wage labor force, in for
    example Bangladesh, South Korea, Taiwan, and Africa, as well as
    ultimately in the ex-Soviet bloc and China. The global proletariat is
    now far larger than ever (which should, surely, put a steely glint of
    hope into every socialist's eye). But it has been radically feminized. It is
    also geographically dispersed, culturally heterogeneous and therefore
    much harder to organize into a united labor movement (cf. Chapter 3).
    Yet it is also living under far more exploitative conditions in aggregate
    than was the case twenty years ago.

3. Global population has also been on the move. The USA now has the
    highest proportion of foreign-born in the country since the 1920s, and
    while there are all sorts of attempts to keep populations out (restrictions
    are far fiercer now than they were, say, in the nineteenth century), the
    flood of migratory movements seems impossible to stop. London,
    Paris, and Rome are far more immigrant cities than they used to be,
    making immigration a far more significant issue worldwide (including
    within the labor movement itself) than has ever been the case before
    (even Tokyo is caught up in the process). By the same token, organizing
    labor or building a coherent oppositional politics to capitalism in the
    face of the considerable ethnic, racial, religious, and cultural diversity
    also poses particular political problems that nation states in general and
    the socialist movement in particular have not found easy to solve.

4. Urbanization ratcheted up into hyper-urbanization, particularly after
    1950 with the pace of urbanization accelerating to create a major
    ecological, political, economic, and social revolution in the spatial
    organization of the world's population. The proportion of an increas-
    ing global population living in cities has doubled in thirty years and we
    now observe massive spatial concentrations of population on a scale
    hitherto regarded as inconceivable. World cities and city systems (as,
    for example, throughout the whole of Europe) have been forming with
    rapidly transforming effects on how the global political economy
    works. The city and the city-region have become far more important
    competitive entities in the world economy, with all sorts of economic
    and political consequences.

5. The territorialization of the world has changed not simply because of
    the end of the Cold War. Perhaps most important has been the

changing role of the state which has lost some (though not all) traditional powers to control the mobility of capital (particularly finance and money capital). State operations have, consequently, been more strongly disciplined by money capital and finance than ever before. Structural adjustment and fiscal austerity have become the name of the game and the state has to some degree been reduced to the role of finding ways to promote a favorable business climate. The 'globalization thesis' here became a powerful ideological tool to beat upon socialists, welfare statists, nationalists, etc. When the British Labour Party was forced to succumb to IMF demands to enforce austerity, it became apparent that there were limits to the national autonomy of fiscal policy (a condition the French also had to acknowledge after 1981). Welfare for the poor has largely been replaced, therefore, by public subventions to capital (Mercedes-Benz recently received one quarter billion dollars of subventions – equivalent to $168,000 subsidy per promised job – in a package from the state of Alabama in order to persuade it to locate there).

But none of this means that the nation state has been 'hollowed out', as writers like Ohmae (1995) claim. To make the contemporary wave of neoliberalism work, the state has to penetrate even more deeply into certain segments of political-economic life and become in some ways even more interventionist than before (Thatcherism was in certain respects highly interventionist). By the same token, the nation state also remains one of the prime defenses against raw market power (as the French have been reasserting since 1995). It is also one key means to defend ethnic and cultural identities and environmental qualities in the face of time-space compression and global commodification. The nation state is, therefore, the prime locus of that 'backlash' against globalization that appeals to populist nationalism.

Reterritorialization has not stopped at the nation state. Global institutions of management of the economy, environment, and politics have proliferated as have regional blocs (like NAFTA and the European Union) at a supra-national scale, and strong processes of decentralization (sometimes through political movements for regional autonomy or, as in the United States, through an increasing emphasis upon States' rights within the Federal system) are also to be found. New local-global forms of relating have been defined and a major shift has occurred in the scales at which the world economy can be grasped, organized, and managed.

6. But while individual states have lost some of their powers, what I call geopolitical democratization has created new opportunities. It became harder for any core power to exercise discipline over others and easier

for peripheral powers to insert themselves into the capitalist competitive game. Money power is a 'leveller and cynic.' But, as Marx
observes, a powerful antinomy then arises: while qualitatively 'money
had no bounds to its efficacy' the quantitative limits to money in the
hands of individuals (and states) limits or augments their social power
in important ways. Given deregulation of finance, for example, it was
impossible to prevent Japan exercising influence as a major financial
power. States had to become much more concerned with their competitiveness (a sub-theme of the globalization argument that has
become important). Competitive states could do well in global competition – and this often meant low-wage states with strong labor
discipline did better than others. Labor control became, therefore, a
vital ideological issue within the globalization argument, again pushing
traditional socialist arguments onto the defensive. Authoritarian, relatively homogeneous territories organized on corporatist principles –
like Singapore, Hong Kong, and Taiwan – have, ironically, done
relatively well in an era when neoliberalism and market freedoms
supposedly became much more the norm. There were and are, however, limitations to the consequent dispersal of capitalist economic
power among nation states since a central authority armed with
significant political and military powers (in this case the United States)
still provides an umbrella under which selective dispersal of economic
power can proceed.

7. 'Globalization' has seemingly produced a new set of global environmental and political problems. I say 'seemingly' because it is not
entirely clear as to whether the problems themselves are new or it is
more a matter that we have become increasingly aware of them through
globalization itself. Widespread and episodically intense fears, for
example, of social disruption arising from the imbalance between
population and resources – the Malthusian specter as it were – have
long been with us. But there has been a growing sense, particularly
since the publication of the celebrated Club of Rome report on *The
Limits to Growth* in 1972, that the flexibility formerly conferred by
having several open frontiers for economic development, migration,
resource extraction, and the construction of pollution sinks was running out. Rapid global population growth, escalating pollution and
waste generation, environmental degradation, and a form of economic
growth that was rather profligate if not downright destructive with
respect to the use of non-renewable as well as renewable resources have
created a whole series of global concerns. Add to that the recognition
that widespread (sometimes global) ecological consequences could be
produced by small-scale activities (such as the local use of various

pesticides like DDT) or that the burgeoning scale of fossil fuel use has been exacerbating climate change, and that losses of habitats and of biodiversity have been accelerating, and it is clear that the environmental issue will assume prominence in global concerns in ways that have not broadly been experienced before. There is, as it were, a translation of traditional environmental concerns (about, say, clean air and water, landscape conservation, and healthy living environments) from a rather local (often urban or regional) to a more global scale.

8. There is, finally, the thorny problem of the relationship of the basic processes I have outlined to the preservation and production of cultural diversities, of distinctive ways of life, of particular linguistic, religious, and technological circumstances of both non-capitalist and capitalistic modes of production, exchange, and consumption. Again, there has been a long historical geography of cultural interventions, cross-influences, and transfers, but the scale and extent of these flows (judged by the volumes of information flow or the march of millions of tourists to say nothing of special artefacts, commodities, and technologies across the globe) suggests a new phase of cultural interpenetration (characterized by rapidity and volatility) with major consequences for modes of thought and understanding. The problem is thorny, however, since it is far too simplistic to see this as purely a movement towards homogeneity in global culture through market exchange. There are abundant signs of all sorts of counter movements, varying from the marketing of cultural difference as a commodity to intense cultural reactions to the homogenizing influence of global markets and strident assertions of the will to be different or special. There is, of course, nothing drastically new in all of this except perhaps the raw fact that the global market implies that there are hardly any places now left outside of market influences. The re-jigging of the map of the earth's human cultures is proceeding apace. The geographical bounding being striven for and the numerous inventions of tradition that are occurring indicate that this is a dynamic field of human activity which is moving in somewhat unpredictable ways. Yet I think it also undeniable that all of this is being strongly driven (albeit in different directions) under the impulses of capitalist globalization.

4 Signs of the times

Two broad questions can be posed about these trends. While everyone will, I think, concede the quantitative changes that have occurred, what really needs to be debated is whether these quantitative changes are great enough and synergistic enough when taken together to put us in a qualitatively new era of capitalist development demanding a radical revision of our theoretical concepts and our political apparatus (to say nothing of our aspirations). The idea that this is the case is signalled primarily by all the 'posts' that we see around us (e.g. post-industrialism, post-modernism).

So has there been a qualitative transformation wrought on the basis of these quantitative shifts? My own answer is a qualified 'yes' to that question immediately accompanied by the assertion that there has not been any fundamental revolution in the mode of production and its associated social relations and that if there is any real qualitative trend it is towards the reassertion of early nineteenth-century capitalist values coupled with a twenty-first century penchant for pulling everyone (and everything that can be exchanged) into the orbit of capital while rendering large segments of the world's population permanently redundant in relation to the basic dynamics of capital accumulation. This is where the powerful image, conceded and feared by international capital, of contemporary globalization as a 'brakeless train wreaking havoc' comes into play. Or, as a disgruntled conservative like John Gray (1998) remarks, although 'the Utopia of the global free market has not incurred a human cost in the way that communism did,' yet 'over time it may come to rival it in the suffering it inflicts.'

If the argument for a limited qualitative shift has to be taken seriously, then the question is how to reformulate both theory and politics. And it is here that my proposed shift of language from 'globalization' to 'uneven geographical development' has most to offer. Uneven conditions offer abundant opportunities for political organizing and action. But they also pose particular difficulties (how, e.g., to cope with the stresses of cultural diversification or of massive income inequalities between rich and poor regions). Understanding both the potentialities and the difficulties is crucial to the formulation of an adequate politics.

It is at this point, however, that the question of globalization as an explicit geopolitical project must be confronted. There are, in this regard, two main features that work together to give the recent push towards further globalization its distinctive form and tone. The first is the stark fact that globalization is undoubtedly the outcome of a geopolitical crusade waged largely by the United States (with some notable allies, such as Britain during the Thatcher years). This has been, as I shall later argue (see Chapter 5), a utopian crusade which has increasingly come in for criticism from radicals and conservatives alike (cf. the recent left-right alliance in attacking the role of the IMF in regulating world economies). But globalization as a process has since 1945 been US centered. It simply would not have happened in the way it has without the US operating as both a driving force and a supervisory agent of the whole process. And this has also meant a certain confusion between specifically US needs and modes of operation (business methods, corporate cultures, traditions of personal mobility and consumerism, political conceptions of individual rights, the law, and democracy) and global requirements. It is hard not to see that the US has over the years frequently thought locally and acted globally, all too frequently without even knowing it. The answer to the question 'who put globalization on the agenda' is, therefore, capitalist class interests operating through the agency of the US foreign, military, and commercial policy.

But the US would not have been able to impose the forms of globalization that have come down to us without abundant support from a wide variety of quarters and places. Many factions of the capitalist class worldwide were more or less happy to align themselves with US policies and to work within the framework of US military and legal protections. In some instances, where they assumed control over the government, they could throw their support, though often with enough local ingenuity (De Gaulle's France comes here to mind) to make it seem as if they were resisting the general extension of capitalist social relations as promoted by the United States. In other instances, most notably Japan, globalization was both responded to and managed in such a distinctive way as to create a competitor to the US model of economic advancement. But even in this case there was a broad acceptance of the globalization argument as necessary to national survival. The Japanese case was not unique, however. And in a sense it is important to see how globalization has been constructed in part through a wide variety of agents (particularly the governments of nation states) thinking locally and acting globally in exactly the same way that the US did as the hegemonic power driving the whole process.

The primary significance for the left in all of these changes is that the relatively privileged position of the working classes in the advanced capitalist countries has been much reduced relative to conditions of labor in the rest of the world (this transition is most glaringly seen in the re-emergence of sweatshops as a fundamental form of industrial organization in New York and Los Angeles over the last twenty years). The secondary point is that conditions of life in advanced capitalism have felt the full brunt of the capitalist capacity for 'creative destruction' making for extreme volatility of local, regional, and national economic prospects (this year's boom town becomes next year's depressed region). The neoliberal justification for all this is that the hidden hand of the market will work to the benefit of all, provided there is as little state interference (and they should add – though they usually don't – monopoly power) as possible. The effect is to make the violence and creative destruction of uneven geographical development (through, e.g., geographical reorganization of production) just as widely felt in the traditional heartlands of capitalism as elsewhere, in the midst of an extraordinary technology of affluence and conspicuous consumption that is instantaneously communicated world-wide as one potential set of aspirations. No wonder even the promoters of globalization have to take the condition of backlash seriously. As long-term promoters of globalization like Klaus Schwab and Claude Smadja – organizers of the influential Davos symposium (cited in Friedman, 1996) – put it:

> Economic globalization has entered a critical phase. A mounting backlash against its effects, especially in the industrial democracies, is threatening a disruptive impact on economic activity and social stability in many countries. The mood in these democracies is one of helplessness and anxiety, which helps explain the rise of a new brand of populist politicians. This can easily turn into revolt.

By 1999, Schwab and Smadja (1999) were still urgently looking for ways to give globalization a human face. Appealing to a rhetoric that echoes in certain respects that of the Zapatistas (see below) they write:

> We must demonstrate that globalization is not just a code word for an exclusive focus on shareholder value at the expense of any other considera-tion; that the free flow of goods and capital does not develop to the detriment of the most vulnerable segments of the population and of some accepted social and human standards. We need to devise a way to address the social impact of globalization, which is neither the mechanical expansion of welfare programs nor the fatalistic acceptance that the divide will grow wider between the beneficiaries of globalization and those unable to muster the skills and meet the requirements of integration in the global system.

Or, as John Gray (1998, 207) concludes:

> [W]e stand on the brink not of the era of plenty that free-marketeers project, but a tragic epoch, in which the anarchic market forces and shrinking natural resources drag sovereign states into ever more dangerous rivalries ... Global market competition and technological innovation have interacted to give us an anarchic world economy. Such an economy is bound to be a site for major geo-political conflicts. Thomas Hobbes and Thomas Malthus are better guides to the world that global laissez-faire has created than Adam Smith or Friedrich von Hayek; a world of war and scarcity at least as much as the benevolent harmonies of competition.

That a conservative commentator of this sort should end up in exactly the same analytic position to that derived from Marx (cf. my own *Limits to Capital*, Chapters 12 and 13, or Greider, 1997) is, of course, intriguing.

Powerful currents of reaction to free-market globalization are setting in (cf., for example, the writings of an eminent capitalist speculator/financier like George Soros, 1996).

The socialist movement has to configure how to make use of these revolutionary possibilities. It has to counter multiple right-wing populist nationalisms (like that advocated by Pat Buchanan in the United States) often edged with outright appeals to a localized fascism (Le Pen in France or the Lombardy Leagues in Italy). It has at the very minimum to focus on the construction of a socially just and ecologically sensitive alternative society. To do this effectively, however, it must come to terms with the conditions of globalization as they currently exist and the growing chorus of demands to reform and manage it. Above all, it must learn to ride the powerful waves of uneven geographical development that make grass-roots and popular organizing so precarious and so difficult. If workers of all countries are to unite to combat the globalization of the bourgeoisie (cf. Chapter 3) so ways must be found to be just as flexible over space in both theory and political practice as the capitalist class has proven to be.

There is one useful way to begin to think of this. Ask first: where is anti-capitalist struggle to be found? The answer is everywhere. There is not a region in the world where manifestations of anger and discontent with the capitalist system cannot be found. In some places or among some segments of a population, anti-capitalist movements are strongly implanted. Localized 'militant particularisms' (I return to the phrase of Raymond Williams) are everywhere to be found, from the militia movements in the Michigan woods (much of it violently anti-capitalist and anti-corporate as well as racist and exclusionary) through the movements in countries like Mexico, India, and Brazil militating against World Bank development projects to the innumerable 'IMF riots' that have occurred throughout the world. And there is plenty of class struggle at work even in the heartlands of capitalist accumulation (such as the extraordinary outbursts of militancy in France in the Fall of 1995 and the victorious UPS workers strike in the United States in 1997).

The interstices of the uneven geographical development hide a veritable ferment of opposition. But this opposition, though militant, often remains particularist (sometimes extremely so) and often threatens to coalesce around exclusionary and populist-nationalist political movements. To say the opposition is anti-capitalist, therefore, is not to say it is necessarily pro-socialist. Broad-based anti-capitalist sentiments lack coherent organization and expression. The moves of one oppositional or protest movement can confound and sometimes check those of another, making it far too easy for capitalist class interests to divide and rule their opposition.

One of the historical strengths of Marxism has been its commitment to synthesize diverse struggles with divergent and multiple aims into a more universal anti-capitalist movement. The Marxist tradition here has an immense contribution to make because it has pioneered the tools with which to find the commonalities within multiplicities and differences (even if it has, at times, submerged the latter rather too readily in the former). The work of synthesis of the multiple struggles that currently exist has to be on-going since the fields and terrains on which these struggles occur and the issues they broach are perpetually changing as the capitalist dynamic and associated global conditions change. Raymond Williams's phrase concerning 'the defense and advancement of certain particular interests, *properly brought together*,' to ground 'a general interest' indicates, then, the core task to be addressed. From the inspiration of that tradition I shall try to distill a number of arguments that seem particularly applicable to the current conjuncture.

CHAPTER 5

Uneven geographical developments and universal rights

On January 30th, 1996 the Zapatista Army for National Liberation in Chiapas, Mexico, issued a call for 'A World Gathering against Neoliberalism and for Humanity.' They proposed a whole series of intercontinental congresses of those opposed to neoliberal capitalism through globalization. Their call (which was partially answered in various ways) pointed out how the power of money everywhere 'humiliates dignities, insults honesties and assassinates hopes.' Renamed as Neoliberalism, 'the historic crime in the concentration of privileges, wealth and impunities democratizes misery and hopelessness.' The name 'globalization' signifies, they suggested, the 'modern war' of capital 'which assassinates and forgets.' Instead of humanity, this neoliberalism 'offers us stock market value indexes, instead of dignity it offers us globalization of misery, instead of hope it offers us emptiness, instead of life it offers us the international of terror.' Against this international of terror, they concluded, 'we must raise the international of hope.' If only everyone touched by the violence of neoliberal globalization could come together politically, then the days of what even advocates of globalization have come to recognize as a 'brakeless train wreaking havoc' would be numbered.

The Zapatista case has fascinated the left for a variety of good and bad reasons. To begin with, the Zapatistas used modern means of communication to good effect. They gave prominence to the way movements can use the internet as a mobilizing tool to promote international campaigns on labor rights, human rights, women's rights, environmental justice, and the like. Some of these movements have experienced considerable success. The anti-Nike campaign over conditions of labor in Nike factories worldwide (particularly in Indonesia and Vietnam) is just one example, but the 1997–8 campaign (led by environmental groups) that challenged and fought off the Multilateral Agreement on Investment (which would have legally empowered multinational capital over state entities with respect to the conditions of foreign investment) illustrated how the global plans of the multinationals could effectively be countered through global oppositions.

73

The other virtue of the Zapatista campaign was precisely the way in which it insisted upon the commonality of the causes that lay behind the distresses being rebelled against. The fact that the uprising occurred on the day that NAFTA took effect symbolized that it was the effects of free trade – the magical mantra of neoliberal rhetoric or, as the *Manifesto* has it, 'that single unconscionable freedom' – that lay at the root of the problems. Furthermore, it appealed fundamentally to notions of the rights and dignity of labor, of indigenous and regional ways of life, in the face of the homogenizing forces of commodification backed by state power.

But on the negative side the reception of the Zapatista movement has unquestionably been characterized by a certain 'romance' of marginality, of a supposedly 'authentic otherness' outside the all-encompassing forms of globalization felt to surround and corrupt oppositional forces at every turn within the heartlands of capitalism. The Zapatista movement thus fell within the orbit of a wide variety of similar movements, such as the Chipko in Nepal, Chico Mendes and the rubber tappers in Amazonia, or Native Americans in the United States, who gained general attention by laying claim to their own cultural identities. They appeared as 'authentic' bearers of a 'true' alternative to a homogenizing and globalizing capitalism. To depict the general reaction on the left in bourgeois democracies this way is not to argue that struggles to sustain or protect distinctive cultural identities are irrelevant or politically vacuous – far from it. Nor does it deny that the experience of marginalization is one of the key resources upon which to build significant movements towards a more universal politics. Indeed, the destructiveness of neoliberal utopianism towards distinctive cultural forms and institutions is, as even a conservative like John Gray argues, one of its most signal failings. But constructing a universal political response purely in such terms falls precisely into the trap of separating off 'culture' from 'political economy' and rejecting the globalism and universality of the latter for the essentialism, specificity, and particularity of the former.

What is missing here is an understanding of the forces constructing historical-geographical legacies, cultural forms, and distinctive ways of life – forces that are omnipresent within but not confined to the long history of capitalist commodity culture and its spatio-temporal dynamics. The Zapatistas, for example, have been deeply touched by colonizing and capitalistic processes over a long time. Their movement and demands for enlightenment values such as 'dignity' – a concept prominent in the 1948 UN Declaration on human rights – is as much a product of history as it is a reflection of circumstance.

I wrap together the forces that work to create and sustain such particular diversities under the rubric of a general theory of uneven geographical developments. So how can such a theory best be depicted?

1 Towards a theory of uneven geographical developments

The argument that follows has two fundamental components. The first concerns 'the production of spatial scale' and the second 'the production of geographical difference.' Let me outline them in general terms.

The production of spatial scales

Human beings have typically produced a nested hierarchy of spatial scales within which to organize their activities and understand their world. Households, communities, and nations are obvious examples of contemporary organizational forms that exist at different scales. We immediately intuit in today's world that matters look differently when analyzed at global, continental, national, regional, local, or household/personal scales. What appears significant or makes sense at one scale does not automatically register at another. Yet we also know that what happens at one scale cannot be understood outside of the nested relationships that exist across a hierarchy of scales – personal behaviors (e.g. driving cars) produce (when aggregated) local and regional effects that culminate in continent-wide problems of, say, acid deposition or global warming. Such an intuitive breakdown is inadequate, however, because it makes it appear as if the scales are immutable or even wholly natural, rather than systemic products of changing technologies, modes of human organization and political struggle.

This does not mean that the relevant scales are defined outside of so-called 'natural' components or influences. Ecological processes and the multiple physical processes that regulate the conditions of lands, waters, and atmospheres themselves operate at a variety of scales (and are usually so represented in the physical and ecological sciences). The definition of where an 'ecosystem' might begin and end and what kind of 'entity' it might be at what kind of scale (a pond or a continent?) is fundamental to the whole question of how to formulate an ecologically sensitive politics. It is, therefore, through a dynamic interaction with what might be called 'natural process' scalars that human beings produce and instantiate their own scales for pursuing their own goals and organizing their collective behaviors.

Consider an example we have already briefly noted. The case of changing territorialization clearly shows that there is nothing 'natural' about political boundaries even if natural features have often played some kind of role in their definition. Territorialization is, in the end, an outcome of political struggles and decisions made in a context of technological and political-economic conditions. The formation of the European Union (a

long process that began with the Monet plan of 1948) is a long case history
of a process of transformation of territoriality from one scale to another.
But changes at this scale have implications elsewhere. There is, for
example, a contemporary debate in France as to how local governments
should be construed (both in territorial organization and powers). A
political cleavage exists between a conception that runs 'localities, departments, nation state' (with all of its strong traditional appeals) versus
'collectivities, regions, Europe' (which reflects the new realities being
forged at the level of the European Union). The outcome of this political
struggle will have important implications for how people can organize their
communal life. In fact the changing powers of local and metropolitan
governments in relationship to nation states and global forces (I think of
everything from inter-urban competition for multinational investment and
'urban entrepreneurialism' to the Agenda 21 element of the Rio agreements which mandated a whole series of local government actions to
contain global warming) has been one significant way in which a particular
scale of human organization has enhanced its role in the last twenty years
(see, e.g., Borja and Castells, 1997).

The scales at which human activity can be orchestrated depend heavily,
of course, on technological innovations (the transport and communications system being vital) as well as upon changing political economic
conditions (trade, geopolitical rivalries and alliances, etc.). They are also
outcomes of class and other forms of political/social struggle at the same
time as they then define the scales at which class struggle must be fought
(see Herod, 1998). As Swyngedouw (1997, 141) argues:

> Spatial scales are never fixed, but are perpetually redefined, contested, and
> restructured in terms of their extent, content, relative importance, and
> interrelations. For example, the present struggle over whether the scale of
> social, labor, environmental and monetary regulation within the European
> Union should be local, national, or European indicates how particular
> geographical scales of regulation are perpetually contested and trans-
> formed. Clearly, relative social power positions will vary considerably
> depending on who controls what at which scale.

When, to take another example, city governments assumed too much
oppositional power in relation to capital accumulation in the Progressive
era in the United States, then the bourgeoisie moved to a different scale and
called for the centralization of powers within the Federal Government
which it was in a better position to control (Margaret Thatcher disbanded
the Marxist-led Greater London Council and reorganized local government for exactly the same reason). The relocation of legal powers to
international organizations like the World Trade Organization with its

far more insidious counterpart of the proposed Multilateral Agreement on Investment, as well as the formation of larger-scale entities like the European Union and NAFTA, have likewise been politically led by capitalist class interests. Oppositional forces often push in the other direction. Secessionist movements and demands for local autonomy (such as those voiced by the Zapatistas) spring up to protect, for example, ethnic minorities, achieved standards of living and welfare protections, or environmental values (where the slogan 'small is beautiful' has considerable purchase).

Even when we regard a particular scale, say that of the urban, as fixed, it turns out that it is also shifting dramatically over time. The scale at which a city/place like Baltimore gets defined (see Plate 3.1) makes it a quite different entity today than was the case 200 years ago. The implications for politics and economy, for sociality and for the meaning that can possibly be put upon the idea of the city (recalling Plato's view that the ideal republican city should have no more than 5,000 people) are legion.

From all these standpoints, therefore, we can meaningfully talk of 'the production of scale' in human affairs and cast it as one vital aspect of any theory of uneven geographical developments (see Smith, 1990; 1992). Plainly, the hierarchical scales at which human activities are now being organized are different from, say, thirty years ago. 'Globalization' in part signifies an important aspect of that shift.

The production of geographical difference

Examination of the world at any one particular scale immediately reveals a whole series of effects and processes producing geographical differences in ways of life, standards of living, resource uses, relations to the environment, and cultural and political forms. The long historical geography of human occupancy of the earth's surface and the distinctive evolution of social forms (languages, political institutions, and religious values and beliefs) embedded in places with distinctive qualities has produced an extraordinary geographical mosaic of socio-ecological environments and ways of life. This mosaic is itself a 'palimpsest' – made up of historical accretions of partial legacies superimposed in multiple layers upon each other, like the different architectural contributions from different periods layered into the built environments of contemporary cities of ancient origin. Some of the layers have greater salience than others (one thinks of the Battle of the Boyne and the separation between Catholics and Protestants in Northern Ireland that continues to be of such immense importance in the region to this day). This geographical mosaic is a time-deepened creation of multiple human activities.

But geographical differences are much more than mere historical-geographical legacies. They are perpetually being reproduced, sustained, undermined, and reconfigured by political-economic and socio-ecological processes occurring in the present. It is just as important to consider how geographical differences are being produced in the here and now as it is to dwell upon the historical-geographical raw materials handed down from previous rounds of activity. Speculators (with international financial backing) seeking to maximize gains from increases in land rent, for example, are now radically re-shaping metropolitan environments in Shanghai and Moscow as well as in London and New York. The more generalized search for differential rent creates geographical differences in the intensity of capital investments, often ensuring that capital-rich regions grow richer while capital-poor regions grow relatively poorer. The differentiating processes are, however, as much ecological and social as purely economic. The development of 'lifestyle niches' and 'communities of shared values' (everything from gay culture districts, religious settlements, and the communes of ecologists to wilderness and wildlife habitat preservation programs pushed by conservation groups) have considerable importance. And ecological changes occurring in their own right and at specific scales (though often with strong anthropogenic inputs, as in the silting of channels, hurricane and flood damages, global warming) also play a role.

These changes have, however, become much more volatile in recent times, partially because of the qualititative shifts that have occurred within the process of globalization itself. The powerful currents of deindustrialization and relocation of manufacturing activities that have swept across the world since 1965 illustrate the speed with which geographical differentiations in manufacturing and employments are now being reconfigured. Economic power has similarly shifted around from one part of the capitalist world to another (flowing easily from the United States to the Pacific Rim and South-East Asia and now back again). It is important, then, to appreciate the volatility and dynamism of contemporary geographical forms. Whole cities and metropolitan regions (think of Seoul or even a long-established city like Barcelona) have been reconfigured and geographically transformed in the space of a generation. Cultural changes (particularly those encouraged by transnational communications systems) are also exhibiting an extraordinary efflorescence and volatility. And there is a good deal of evidence (accompanied by a good deal of worry) that ecological and natural systems are in accelerated states of change (with a good deal of attendant stress). So even though geographical variations reflect and incorporate material, historical, cultural, and political legacies from the past, it is a gross error to presume they are even relatively static let alone immutable. The geographical mosaic has always been in motion at all scales. Its contemporary volatility leaves the widespread but surficial impression, however, of global anarchy rather than of the working out of systemic forces of production of uneven geographical developments.

The analytics and politics of uneven geographical developments

The general conception of uneven geographical development that I have in mind entails a fusion of these two elements of changing scales and the production of geographical differences. We need to think, therefore, about differentiations, interactions, and relations across and within scales. A common error of both analytical understanding and political action arises because we all too often lock ourselves into one and only one scale of thinking, treating the differences at that scale as *the* fundamental line of political cleavage. This is, I submit, one of the most pervasive errors to arise from all the globalization talk to which we are now exposed. It erroneously holds that everything is fundamentally determined at the global scale.

This is such a serious issue that it warrants brief elaboration. Consider, for example, Huntington's (1996) view that the future depends on some great clash of civilizations – East versus West. He assembles a lot of history and ideas to make what seems like a plausible argument. But in the end he fails to make proper sense even at that global scale precisely because he does not recognize the imbrication of other scales of activity which run counter to his simplifying thesis. Different actors and agents often operate (sometimes craftily) across different scales. A diaspora of Chinese business entrepreneurs (armed with a whole set of values drawn from their own civilization's history) operates capitalistically (armed with 'universalizing' Western concepts about wealth, power, and technology) and globally in a world of scattered but tightly bound ethnically distinct Chinese communities. Forming enclaves often isolated from the habits and laws of the host country, these communities provide pools of captive and compliant labor for Chinese entrepreneurs, giving the latter a certain competitive advantage. The distinctive globalization achieved by this diaspora rests, therefore, upon the existence of distinctive communities defined at a much more local scale. At that scale, Chinatown in New York looks like a community concerned to retain cultural distinctness and identity (and we outsiders might sympathize politically with that in the same way we enjoy the restaurants). But this misses out entirely on the other highly exploitative and often illegal relation of labor practices set up in Chinatown under the control of the global diaspora of Chinese entrepreneurs. The latter also make it their business to promote the exclusionary qualities of this localized ethnic enclave (with its own newspapers and cultural forums) for purposes of labor control. Where, to revert to Huntington's categories, is the East and the West in this example?

It is hard to understand what happens in such communities as China-town without considering how processes operating primarily at other scales (transnational migratory currents, interest rate fluctuations, shifts in terms of trade, relocations of industrial activity, environmental changes, etc.) impinge upon them. It is not hard to multiply examples of this sort though with different ramifications (how the international bank known as BCICC was formed and collapsed is a wonderful example of how intrigue at all sorts of different spatial scales built into a momentous event that had widespread economic, social, and political ramifications at all scales). Conversely, it is precisely the ability of the Zapatistas to transform what is in effect a local struggle with particular issues (some of which are hard to appreciate on the outside) onto a completely different scale of analytics and politics that has made the uprising so visible and so politically interesting.

The upshot is to render all ways of thinking that operate only at one scale at least questionable if not downright misleading. But it then becomes conceptually difficult to work simultaneously with multiplying and volatile geographical differentiations operating at scalars which are themselves rapidly changing. The problems become highly complicated and hard to unravel in their details. Fortunately, there is now an extensive and instructive literature on how, for example, to relate the local to the global or, in Swyngedouw's (1997) somewhat ungainly terminology, analyze the phenomena of 'glocalization.' Simplifications are inevitably called for. But it is vital to retain an eternal vigilance. Uneven geographical developments work to generate all sorts of unintended consequences for both capitalistic and socialistic projects alike. Furthermore, the habit of interpreting everything through the lens of conflict measured at one scale and then acting on those interpretations can have disastrous consequences (this was one of the most deleterious aspects of that 'cold war mentality' that had the US pursue its geopolitical crusade for globalization by interpreting every struggle for social justice everywhere in pro- or anti-communistic terms).

Uneven geographical developments of the sort I have outlined plainly pose serious barriers to the 'proper bringing together' of multiple parti-cular interests into some framework expressive of the general interest. There are many conflicts, for example, where local concerns over access to resources, opening up better life chances and gaining elementary forms of economic security outweigh all efforts to cultivate respect for global concerns about such important issues as, say, human rights, greenhouse gas emissions, preservation of biodiversity, or the regulation of land uses to prevent deforestation or desertification.

Yet it is also through an understanding of uneven geographical developments that we can more fully appreciate the intense contradictions that now exist within capitalist trajectories of globalization. This helps redefine possible fields of political action. Globalization entails, for example, a great deal of self-destruction, devaluation and bankruptcy at different scales and in different locations. It renders whole populations selectively vulnerable to the violence of down-sizing, unemployment, collapse of services, degradation in living standards, and loss of resources and environmental qualities. It puts existing political and legal institutions as well as whole cultural configurations and ways of life at risk and it does so at a variety of spatial scales. It does all this at the same time as it concentrates wealth and power and further political-economic opportunities in a few selective locations and within a few restricted strata of the population.

The positive and negative effects vary in intensity from place to place. It is important to recall, therefore, that globalization has always been a specific project pursued and endorsed by particular powers in particular places that have sought and gained incredible benefits and augmentations of their wealth and power from freedoms of trade. But it is precisely in such localized contexts that the million and one oppositions to capitalist globalization also form, crying out for some way to be articulated as a general oppositional interest. This requires us to go beyond the particularities and to emphasize the *pattern* and the systemic qualities of the damage being wrought across geographical scales and differences. The pattern is then describable as the uneven geographical consequences of the neoliberal form of globalization.

The analysis can then be broadened outwards to embrace a diverse array of social and environmental issues that on the surface appear disconnected. Issues like AIDS (a devastating problem in Africa where, in some countries, as many as one in four are HIV positive), global warming, local environmental degradation, and the destructions of local cultural traditions can be understood as inherently class issues. Building some sort of international community in class struggle can better alleviate the conditions of oppression across a broad spectrum of socio-ecological action. This is not, I emphasize, a plea for pluralism, but a plea that we seek to uncover the class content of a wide array of anti-capitalist concerns. This idea will encounter opposition from within the radical left, for to insist upon a class formulation often invites dismissal as pure sectarianism of the old-guard sort (to say nothing of being rejected as *passé* in the media and academia). But 'all for one and one for all' in anti-capitalist struggle continues to be a vital slogan for any effective political action and that inevitably implies some sort of class politics, however defined.

Such work of synthesis has, however, to re-root itself in the organic conditions of daily life. This does not entail abandoning the abstractions that Marx and the Marxists have bequeathed us, but it does mean re-validating and revaluing those abstractions through immersion in popular struggles at a variety of scales, some of which may not appear on the surface to be proletarian in the sense traditionally given to that term. In this regard, Marxism has its own sclerotic tendencies to combat, its own embedded fixed capital of concepts, institutions, practices, and politics which can function on the one hand as an excellent resource and on the other as a dogmatic barrier to action. We need to discern what is useful and what is not in this fixed capital of our intellect and politics and it would be surprising if there were not, from time to time, bitter argument on what to jettison and what to hold. Nevertheless, the discussion must be launched (see Part 4).

For example, the traditional Marxist categories – imperialism, colonialism, and neocolonialism – appear far too simplistic to capture the intricacies of uneven spatiotemporal developments as these now exist. Perhaps they were always so, but the reterritorialization and re-spatialization of capitalism particularly over the last thirty years makes such categories seem far too crude to express the geopolitical complexities within which class struggle must now unfold. While a term like 'globalization' repeats that error in a disempowering way for socialist and anti-capitalist movements, we cannot recapture the political initiative by reversion to a rhetoric of imperialism and neocolonialism, however superior the political content of those last terms might be. Here, too, I believe a shift to a conception of uneven geographical developments can be helpful in order both to appreciate the tasks to be surmounted and the political potentialities inherent in multiple militant particularist movements of opposition that cry out to be combined.

Consideration of the scale problem draws our attention back to the 'proper' scale at which oppositions may be formulated. The grand (and in my view unfortunate) divisions between anarchists and Marxists over the years has in part been over the appropriate scale at which oppositions should be mounted and the scale at which an alternative social form should be envisaged and constructed (see Forman, 1998, and Thomas, 1985). Within the Marxist movements there are similar divisions between, for example, Trotskyists who tend to believe revolution must be global or nothing, or more traditional communist parties that believe the conquest of state power is the most important immediate (and *de facto* sole) objective worthy of discussion. All of these divisions, it seems to me, would be better understood and more clearly articulated through the formulation of a theory of uneven geographical developments. The potential unities between them (as opposed to their often bitter differentiations) would then be seen as a problem of how to connect political activities across a variety of geographical scales.

In this regard there is one further organizational point of considerable importance. The traditional method of Marxist intervention has been via an avant-garde political party. But difficulties have arisen from the superimposition of a single aim, a singular objective, a simple goal formulated usually at one specific scale upon anti-capitalist movements that have a multiplicity of objectives and scalars. As many critics within the Marxist tradition have pointed out, the emancipatory thrust of Marxism here creates the danger of its own negation. It suppresses and even represses at one scale in order to achieve at another. It is therefore vital to understand that liberating humanity for its own development is to open up the production of scales and of differences, even open up a terrain for contestation within and between differences and scales, rather than to suppress them. This is something that the right wing sometimes argues for – though it rarely practices it as its turn to religious fundamentalism indicates. But we should note the power of the argument. The production of real (by which I mean affective and socially embedded) as opposed to commodified cultural divergence, for example, can just as easily be posed as an aim of anti-capitalist struggle. The aim to create a unified, homogeneous socialist person was never plausible. After all, capitalism has been a hegemonic force for the production of a relatively homogeneous capitalist person and this reductionism of all beings and all cultural differences to a common commodified form of uneven geographical development has itself generated strong anti-capitalistic sentiments. The socialist cause must, surely, be just as much about emancipation from that bland homogeneity as it is about the creation of some analogous condition. This is not, however, a plea for an unchecked relativism or unconstrained postmodern eclecticism, but for a serious discussion of the relations between commonality/difference, the particularity of the one and the universalism of the other. And it is at this point that socialism as an alternative vision of how society will work, social relations unfold, and human potentialities be realized, itself becomes the focus of conceptual work. Uneven geographical developments need to be liberated from their capitalistic chains.

2 Political universals and global claims

It would be an egregious error to conflate the 'global' of globalization with more general claims to universality (of truths, moral precepts, ethics, or of rights). Yet it would also be wrong to miss the rather powerful connexion between them. I want, therefore, to look more closely at this relation. I shall argue for a certain kind of universalism within which uneven geographical advancements of human interests might flourish in more interesting and productive ways. The proper establishment of conditions at one scale – in this case the global and the universal taken together – is here seen as a necessary (though not sufficient) condition to create political and economic alternatives at another.

The recent phase of globalization has posed a whole series of open questions concerning universality. This has been so because globalization has forced us to consider, in political rhetoric and to some degree in terms of political-economic fact, the nature of our 'species being' on planet earth at the same time as it forces us to consider the rules and customs through which we might relate to each other in a global economy where everyone to some degree or other relates to and is dependent upon everyone else. Such conditions superimpose a certain commonality (most particularly evident in environmental debates) upon the whole world of difference and otherness that is the normal grist for political theorizing in postmodern circles. In fact the whole postmodern movement might well be construed as a movement to celebrate or mourn that which is in any case on the brink of disappearing. 'Otherness' and 'difference' (and even the idea of 'culture' itself as a regulatory ideal, as Readings [1996] so trenchantly points out) become important to us precisely because they are of less and less practical relevance within the contemporary political economy of uneven geographical developments even as they come to ground bitter ethnocentric and communitarian violence.

Consider, for example, the contemporary saliency in political debate of what many of us have come to call 'the local/global nexus.' The fascination with 'local knowledges' (a theme made popular particularly by Geertz) and local cultures, with the politics of 'place' and of 'place construction' is everywhere in evidence. It has spawned an immense literature across a variety of disciplines describing local variations in ways of life, in structures of feeling, in forms of knowledge, in modes of social relating and of production, in socio-ecological structures, and in values and beliefs. It has also spawned a normative literature in which some kind of localism or communitarianism (sometimes utopian) is held up as an ideal of social life to which we should aspire, as a proper structure through which all universalizing concepts (such as justice) should be mediated and translated into specifically acceptable local terms.

I have no intention of trying to review or dissect this vast literature here. But I do want to make one crucial point. It can itself at least in part be read as a distinctive product of the recent trends in globalization. Its dominant theses can then be interpreted in three ways:

1. as a throwback or as a lag in which the consciousness (fueled by language, education, discursive regimes, the media, political agitations, etc.) of a past era (e.g. nationalism or patriotism) is preserved in the face of a crying need to adopt what Nussbaum (1996) calls the 'cosmopolitanism' appropriate to today's world;

2. as the focus for a political or even utopian longing that desires a simpler, less volatile, more manageable, and securing life in which affective and direct interpersonal relations established on a local basis are cultivated and enhanced in the face of the seemingly abstract and impersonal forces of globalization; and

3. as a recognition that for most people the terrain of sensuous experience and of affective social relations (which forms the material grounding for consciousness formation and political action) is locally circumscribed by the sheer fact of the material embeddedness of the body and the person in the particular circumstances of a localized life.

There are then three lines of response. The first is to regret the passing of the old order and to call for the restoration of past values (religious, cultural, national solidarities, or whatever). Much of the current thinking on both the left (cf. Greider, 1997) and the right (cf. Gray, 1998) is hopelessly infected with nostalgia in its prescriptions and predictions.

The second is to pursue the utopian vision of some kind of communitarianism (including movements of national redemption as an answer to the alienations and abstractions of a globalizing political economy and culture). Many political movements now trend in this direction, sometimes appealing to some sort of political mythology laced with nostalgia for a lost golden age of organic community.

The third path is to take globalization literally and make universal claims of precisely the sort that the Zapatistas have advanced from their mountainous retreats in Southern Mexico. These claims rest firmly on local experience but operate more dialectically in relation to globalization. They appeal, for example, to the embeddedness of local cultural forms but also use the contradiction implicit in the current acceptance throughout the world of certain norms and ways of 'doing business' and of defining 'freedoms' and the right to choose. Globalization implies widespread (though often informal, grudging, corrupted, and even surficial) acceptance of certain bourgeois notions of law, of rights, of freedoms, and even of moral claims about goodness and virtue. This is the albeit often weak but nevertheless omnipresent political corollary of maintaining an open field for capital investment, accumulation, and labor and resource exploitation across the surface of the earth. This was and still is manifest in the moral dimension to the geopolitical crusade for globalization launched under the umbrella of US power. Keeping the world safe for democracy and freedoms of expression was and is seen as intimately tied to keeping the world safe for capital and vice versa. Here lies the root of a whole series of contemporary paradoxes and contradictions that create opportunities and potentialities for progressive forms of political action.

Consider, for example, the Universal Declaration of Human Rights signed in 1948 as part of the UN Charter. Mainly constructed at the behest of the United States, it was immediately put to use as a tool in the struggles around the Cold War. But it was a rather weak tool, particularly to the degree that the United States itself paid no attention to it in the face of its own political expediency. For this reason, Amnesty International was founded in 1961 as a transnational organization dedicated to raising the issue of universal rights in a geopolitically divided, socially fragmented, but also globalizing world. It is only in the last twenty years, however, that the meaning of these rights have assumed a new saliency (in part with the end of decolonization struggles such as those in Africa and Asia). During the Carter presidency, human rights issues became more visible with significant effects (particularly in Central and South America). With the end of the Cold War, the direct use of this tool for narrow political purposes has become less common (with notable exceptions such as that of China and of Cuba). The question of the application of the Universal Declaration of Human Rights now hovers over the world as a contested set of universal principles looking for application (see Alston, 1992). What, then, do we make of these principles under contemporary conditions of globalization?

We can, of course, take a rigid Marxist stance (pioneered in Marx's attack upon Proudhon's conception of 'eternal justice') that says all conceptions of rights are captive to bourgeois institutions and therefore to build any politics around such claims is pure reformism. Or we can take a broadly postmodern stance and attack them as a mere stepchild of erroneous patterns of Enlightenment thought, incapable of adaptation to a world of incommunicability and irreconcilable cultural difference. And considerable evidence has been assembled to show how the claim of universality can all too easily become a vehicle for repression and domination of other interests and that the claim is in any case far too deeply embedded in the concept of eighteenth-century liberalism to ever be mobilized for any other or deeper emancipatory purpose.

The alternative is to recognize that all claims to universality are fraught with difficulty and that the distinction between reformism and revolution is never as sharp as some Marxists maintain. Nor can the distinction between particularity and universality be so easily defined. The problem is then to find ways to broaden and amplify the scope of human rights in ways that are as sympathetic as possible to the right to be different or the 'right to the production of space.' Any strict and narrow interpretation of human rights must be contested. But in practice we observe that such contestation has been a relatively permanent feature over the past fifty years as the meaning and application of the UN principles have had to be fought out on the ground from place to place and from case to case (cf. Alston, 1992; Phillips and Rosas, 1995), even leaving some of the stated principles of 1948 a practical dead letter while in other instances exhuming largely ignored rights (such as those of minorities) in order to confront new problems (such as those encountered in what was once Yugoslavia and the Soviet Union).

Contestation existed, of course, at the very beginning. In 1947, for example, the Executive Board of the American Anthropological Association submitted a collective statement to the United Nations Commission on Human Rights. It began by pointing out that, in the world order, 'respect for the cultures of differing human groups' was just as important as respect for the personal individual. Respect for the latter must construe the individual 'as a member of the social group of which he is a part, whose sanctioned modes of life shape his behavior, and with whose fate his own is thus inextricably bound.' It then went on to worry that the UN Declaration would appear as 'a statement of rights conceived only in terms of the values prevalent in the countries of Western Europe and America.'

In seeking to ameliorate this difficulty, the Executive Board insisted that any statement of rights must recognize how 'values' and 'desires' are arrived at through subtle processes of learning in cultural settings which differ markedly. While tolerance between different accepted belief systems was possible:

> In the history of Western Europe and America ... economic expansion, control of armaments, and an evangelical religious tradition have translated the recognition of cultural differences into a summons to action. This has been emphasized by philosophical systems that have stressed absolutes in the realm of values and ends. Definitions of freedom, concepts of the nature of human rights, and the like, have thus been narrowly drawn. Alternatives have been decried, and suppressed where controls have been established over non-European peoples. The hard core of *similarities* between cultures has consistently been overlooked.
>
> The consequences of this point of view have been disastrous for mankind ...

The Executive Board then went on to advance three fundamental propositions:

1. The individual realizes his personality through his culture, hence respect for individual differences entails a respect for cultural differences.

2. Respect for differences between cultures is validated by the scientific fact that no technique of qualitatively evaluating cultures has been discovered.

3. Standards and rules are relative to the culture from which they derive so that any attempt to formulate postulates that grow out of the beliefs or moral codes of one culture must to that extent detract from the applicability of any Declaration of Human Rights to mankind as a whole.

Cultural relativism and critical commentary on univeralism have long been features of political and academic thought. But far from advocating a 'formless relativism,' as some might now depict or even prefer it, the Executive Board drew attention to that 'hard core of *similarities* between cultures' that has been 'consistently overlooked' (the italicized 'similarities' is in the original) and accepted the idea that 'world wide standards of freedom and justice' must 'be basic.' The problem, then as now, is to devise an effective world order which 'permits the free play of personality of the members of its constituent social units, and draws strength from the enrichment to be derived from the interplay of varying personalities.' The Executive Board concluded:

> Only when a statement of the right of men to live in terms of their own traditions is incorporated into the proposed Declaration, then, can the next step of defining the rights and duties of human groups as regards each other be set upon the firm foundation of the presentday scientific knowledge of Man.

These are, it seems to me, exactly the kinds of claims that the Zapatistas are making. On the one hand they appeal repeatedly and powerfully to the concept of 'dignity' and the universal right to be treated with respect. On the other they make claims based on locality, embeddedness, and cultural history which emphasize their unique and particular standing as a socio-ecological group. Universality and particularity are here dialectically combined.

In practice, of course, the whole field of application of human rights since 1948 has been dominated by an interpretive separation between civil and political rights on the one hand and economic, social, and cultural rights on the other. The latter set of rights has, until recently, been kept off limits to discussion, even though it is actually present within the 1948 Declaration. One of the impacts of globalization and the rise of many multinational and transnational forms of capital is to make it harder and harder to sustain such a strict separation. The question of economic rights begins to loom large upon the agenda of any restatement of universal rights. There are several signs of emergent forms of class struggle that have precisely as their objective the bringing of economic rights within the general domain of consideration of human rights. The struggle to make transnational institutions accountable with respect to such issues is by now a familiar objective. The most spectacular example was the public reaction to the execution of Ken Saro-Wiwa and his eight co-defendants in Nigeria for defending the rights of the Ogoni people in the face of Shell's exploitation of the local oil reserves. Shell could not easily sustain its fictitious position that it cannot interfere in the internal political affairs of a country like Nigeria (it patently does so all the time) and was forced ultimately to concede limited support to some of the main tenets of human rights law (though, sadly, only years after the executions rather than before). It is hard, of course, to go beyond the spectacular case but the vulnerability of transnational institutions to such questions is becoming apparent.

There are now much broader struggles in the offing that echo some of the national struggles that occurred in the nineteenth century. International conventions to outlaw child labor are proposed, and international accords to circumscribe and regulate sweatshops worldwide are being debated. There is now an international movement to push for a global 'living wage' as a standard to be observed across all industries and in all places. Such movements are growing in significance (Pollin and Luce, 1998), even though they remain small and under-reported in the mainstream media (though not on the internet). If, in the broad terms of capitalism's logic, a certain kind of democratization and capital accumulation go hand in hand, then how can economic rights be kept apart from human and civil rights more generally? This connection is already present in the 1948 Declaration. Consider, for example, Articles 22 to 25:

Article 22
Everyone, as a member of society, has the right to social security and is entitled to realization, through national effort and international cooperation and in accordance with the organization and resources of each State, of the economic, social and cultural rights indispensable for his dignity and the free development of his personality.
Article 23
1. Everyone has the right to work, to free choice of employment, to just and favourable conditions of work and to protection against unemployment.
2. Everyone, without any discrimination, has the right to equal pay for equal work.
3. Everyone who works has the right for just and favourable remuneration ensuring himself and his family an existence worthy of human dignity, and supplemented, if necessary, by other means of social protection.

4. Everyone has the right to form and to join trade unions for the protection
of his interests.

Article 24

Everyone has the right to rest and leisure, including reasonable limitation of
working hours and periodic holidays with pay.

Article 25

1. Everyone has the right to a standard of living adequate for the health and
well-being of himself and of his family, including food, clothing, housing
and medical care and necessary social services, and the right to security
in the event of unemployment, sickness, disability, widowhood, old age
or other lack of livelihood in circumstances beyond his control.

2. Motherhood and childhood are entitled to special care and assistance.
All children, whether born in or out of wedlock, shall enjoy the same
social protection.

What is striking about these articles (particularly when stripped of their
strong gender bias – itself an easy indicator as to how fraught all universal
declarations are) is the degree to which hardly any attention has been paid
over the last fifty years to their implementation or application and how
almost all countries that were signatories to the Universal Declaration are
in gross violation of these articles. Strict enforcement of such rights
would entail massive and in some senses revolutionary transformations in
the political-economy of capitalism. Neoliberalism could easily be cast,
for example, as a gross violation of human rights. Certainly, the whole
policy trajectory over the past twenty years in the United States (continued most dramatically through the reform of welfare in the Clinton
administration) has been diametrically opposed to the securing of such
rights.

Practical applications of human rights have also typically distinguished
between rights in the public and private spheres. The former (like voting
and freedom of political expression) have been highlighted whereas the
latter (the subservience of women within the family, cultural practices
like genital mutilation, and the rights of women to control their own
bodies or resist domestic violence) are broadly ignored. The result has
been an intense gender bias within the concept of human rights (captured
within the language of the original articles). Here, too, the neoliberal
sweeping away of many distinctions between private and public through,
for example, privatization of many formerly public functions and, conversely, the bringing of many supposedly private issues (over, say,
reproductive rights and personal health) into the public domain has made
it harder and harder to preserve the distinction. The broadly regretted
loss of a clearly demarcated 'public sphere' for politics here creates an
opportunity to redefine notions of human rights in general.

At the outset, I considered the politics of globalization mainly in its negative and disempowering aspects. What now becomes clear is that the contradictions and paradoxes of globalization offer opportunities for an alternative progressive politics. Contemporary globalization offers a rather special and unique set of conditions for radical change.

There is, first of all, the widespread demand for reform of the system given its manifest instabilities (the periodic financial difficulties, phases of deindustrialization, and the like) and deepening economic inequalities. Secondly, the environmental difficulties are everywhere apparent and many of these also call for regulatory action and interventions at all scales including that of the global. New institutional arrangements are therefore widely considered as necessary to guarantee more stable and more sustainable economic growth. Thirdly, the spread of Western ways of thinking about self-fulfillment and self-realization (as represented by the 1948 Declaration) has unleashed a set of powerful forces of mounting economic, social, and cultural frustration. The move towards a more universal popular culture, though it may be regretted, pushes demands for self-fulfillment while feeding alienation and frustration. The global crusade to impose bourgeois democratic rights and freedoms has also promoted a ground-swell of global cultural revolution that is anti-authoritarian, individualistic, subversive of deference, and to some degree awkwardly egalitarian (this is particularly evident in the unfolding of women's movements in, say, Japan and South-East Asia). So while it may be true that local differentiations and adaptations, even local innovations and initiatives, are widespread on questions of rights and values, there is a discernable trend to recognize how these variations are part of some family of meanings rather than incommunicably different.

Globalization, in short, poses the question of our 'species being' on planet earth all over again (see Chapter 10). It opens up terrains both of conceptual and theoretical debate, and of political struggles (shadowy forms of which can already be discerned). Above all it makes it necessary and possible to redefine universal human rights that stretch far beyond those acknowledged in 1948 (a matter I shall take up again in Chapter 12). Such rights will not be freely given or conferred precisely because they may lead towards revolutionary changes in social orders and political economies. They will be achieved only through struggles. This will entail intense and often irreconcilable arguments, particularly when the rights are in contradiction to each other or, more significantly, set precedents that are antagonistic to the workings of market capitalism.

Consider, for example, the general idea that laborers everywhere should be treated with dignity, that they should be accorded a 'living wage' that guarantees them the minimum of economic security and adequate access to life chances. That universal conception, thoroughly consistent with the Universal Declaration of 1948, plainly runs up against the conditions of uneven geographical development that capital has both fed upon and, in many instances, actively produced. It challenges neoliberalism in a fundamental way because it interferes with the functioning of labor markets in a fundamental way. Yet we also know that the demand for a 'living wage' evidently means something different depending upon historical-geographical conditions. A living wage in Dacca or Bombay is not the same as that which might be required in Johannesburg, Duluth, Lulea, or New York City. Does this imply the struggle for a global right to a living wage is impossible or unreasonable?

On July 17th–19th 1998, some forty representatives from worker rights organizations in the United States, the Caribbean, Central America, Mexico, Canada, and Europe met to consider exactly that question (see Benjamin, 1998). They concluded that a campaign for setting a living wage as a global standard (initially in the shoe and apparel industries) was both feasible and worthwhile even though any formula to calculate that wage, no matter how carefully crafted to take account of cultural, social, and economic differences between countries and regions, would doubtless be the focus of intense controversy. It would, moreover, be unlikely to enjoy universal acceptance. But, they concluded, 'the more controversy over this formula and the more alternative formulas proposed the better' (Benjamin, 1998, 4). Simply by drawing the industry and the public into a debate over which formula to adopt forces the issue of a living wage, defined as 'a wage with dignity,' to the forefront of the political agenda in exactly the same way that the adoption of the Universal Declaration of Human Rights in 1948 placed certain issues about universal human rights (however controversial or fuzzy) irrevocably upon the global agenda.

The dilemma to which this points becomes even more starkly evident when we attempt to re-embed (or reconcile) the supposed universality of human rights within a theory of uneven geographical developments. The considerations offered up by the American anthropologists in 1947 point up the problem even though, it must be said, their tendency to fetishize cultures as somehow independent entities was as dubious then as it is obviously inappropriate now. The strength of the UN Declaration lies in the way it brings together the universal and the global scale on the one hand and the microscale of the body and the political person on the other. But it pays scant attention to the various other scales at which meaningful human associations can be constructed (though the nation state is seen as the mediating entity accountable for the guarantee and enforcement of rights). And it also pays scant attention to all those infinite variations of custom and habit, of ways of life and structures of feeling, that anthropologists and geographers have so long focussed upon as crucial aspects of human existence. The right to uneven geographical development, to build different forms of human association characterized by different laws, rules, and customs at a variety of scales appears in this regard as fundamental a human right as any other. The contradictions and tensions implicit in such an argument are readily apparent. The right to difference confronts the universality of rights. In a way this can be viewed as a fortunate rather than desperate condition. For it is precisely out of such unresolvable tensions that new states of human being can be constructed.

Nevertheless, the thorny problem of how to reconcile the right to uneven geographical development (political, economic, and cultural) and some universal ideals about rights will never go away. But to say that it will always be with us is not to construe the dilemma as so intractable as to be beyond any kind of reason. Between the absolutist relativism that says that nothing that happens in, say, Jakarta or Vietnam or even in Boulder and the inner city of Baltimore is a proper subject for my moral or political judgement and the absolutism that rigidly views universality as a matter of total uniformity and equality of judgement and treatment, there is abundant room for negotiation. The recent resurrection of long-standing questions about minority rights, leading to the UN Declaration of 1992 on the subject, is a case in point. The context, as Thornberry (1995) notes, rests on the clash of contradictory trends:

> On the one hand, there is a movement towards the internationalization or globalization of environmental, resource, humanitarian and rights issues which transcends Cold War inhibitions. On the other hand, we are witness to the emergence or recrudescence within and between states of virulent forms of ethnocentrism, hatred of diversity, the exhumation of buried antagonisms and the obscenity of 'ethnic cleansing.' (13)

Recognizing that societies are likely to become more rather than less multicultural and plural in years to come and that cultural differentiation is an on-going process (and that rights therefore must stretch far beyond the 'museumification' of pre-existing cultures), the UN Declaration of 1992 and its subsequent development illustrate one of the ways in which the idea of rights can be extended to deal with problems of uneven geographical development (Phillips and Rosas, 1995). Far from providing a license for the proliferation of interethnic rivalries and hatreds, a properly crafted system of rights may here provide a vehicle 'to safeguard equality between all human beings in society; to promote group diversity where required to ensure everyone's dignity and identity; and to advance stability and peace, both domestically and internationally' (65).

The construction of political forces to engage in such dialogues within some adequate institutional frame then becomes the crucial mediating step in bringing the dialectic of particularities and universalities into play on a world stage characterized by uneven geographical developments. And that, presumably, is what 'the proper bringing together of particular interests' is all about.

Marx was not impressed with talk about rights. He often saw it as an attempt to impose one distinctive set of rights – those defined by the bourgeoisie – as a universal standard to which everyone should aspire. But if workers of the world are to unite then surely it has to be around some conception of their rights as well as of their historical mission? As a matter of practical politics some notion of rights appears indispensable. The First International, much influenced by Marx, appealed after all to 'the simple laws of morals and justice, which ought to govern the relations of private individuals' as the proper basis to conduct 'the intercourse of nations' (Marx and Lenin, 1940, 23). And on questions of rights the bourgeoisie has created such a maelstrom of contradictions on the world stage that it has unwittingly opened up several paths towards a progressive and universalizing politics at the global scale. To turn our backs on such universals at this stage in our history, however fraught or even tainted, is to turn our backs on all manner of prospects for progressive political action. Perhaps the central contradiction of globalization at this point in our history is the way in which it brings to the fore its own nemesis in terms of a fundamental reconception of the universal right for everyone to be treated with dignity and respect as a fully endowed member of our species.

PART 2

ON BODIES AND POLITICAL PERSONS
IN GLOBAL SPACE

CHAPTER 6

The body as an accumulation strategy

[I]t is crystal clear to me that the body is an accumulation strategy in the deepest sense. (Donna Haraway, *Society and Space*, 1995, 510)

Capital circulates, as it were, through the body of the laborer as variable capital and thereby turns the laborer into a mere appendage of the circulation of capital itself. (David Harvey, *The Limits to Capital*, 1982, 157)

In fact the two processes – the accumulation of men and the accumulation of capital – cannot be separated.
(Michel Foucault, *Discipline and Punish*, 1975 [1995], 221)

Why focus on these citations? In part the answer rests on the extraordinary efflorescence of interest in 'the body' as a grounding for all sorts of theoretical enquiries over the last two decades or so. But why this efflorescence? The short answer is that a contemporary loss of confidence in previously established categories has provoked a return to the body as the irreducible basis for understanding (cf. Chapter 1 and Lowe, 1995, 14). But viewing the body as the irreducible locus for the determination of all values, meanings, and significations is not new. It was fundamental to many strains of pre-Socratic philosophy and the idea that 'man' or 'the body' is 'the measure of all things' has had a long and interesting history. For the ancient Greeks, for example, 'measure' went far beyond the idea of comparison with some external standard. It was regarded as 'a form of insight into the essence of everything' perceived through the senses and the mind. Such insight into inner meanings and proportionalities was considered fundamental in achieving a clear perception of the overall realities of the world and, hence, fundamental to living a harmonious and well-ordered life. Our modern views, as Bohm (1983) points out, have lost this subtlety and become relatively gross and mechanical, although some of our terminology (e.g. the notion of 'measure' in music and art) indicates a broader meaning.

The resurrection of interest in the body in contemporary debates does provide, then, a welcome opportunity to reassess the bases (epistemological and ontological) of all forms of enquiry. Feminists and queer theorists have pioneered the way as they have sought to unravel issues of gender and sexuality in theory and political practices. And the question of how measure lost its connexion to bodily well-being has come back into focus as an epistemological problem of some significance (Poovey, 1998). The thesis I want to pursue here is that the *manner* of this return to 'the body as the measure of all things' is crucial to determining how values and meanings are to be constructed and understood. I want in particular to return to a broader relational meaning of the body as 'the measure of all things' and propose a more dialectical way of understanding the body that can better connect discourses on the body with that other discursive shift that has placed 'globalization' at the center of debate.

1 Bodily processes

I begin with two fundamental propositions. The first, drawn from writers as diverse as Marx (1964 edition), Elias (1978), Gramsci (1971 edition), Bourdieu (1984), Stafford (1991), Lefebvre (1991), Haraway (1991), Butler (1993), Grosz (1994), and Martin (1994), is that the body is an unfinished project, historically and geographically malleable in certain ways. It is not, of course, infinitely or even easily malleable and certain of its inherent ('natural' or biologically inherited) qualities cannot be erased. But the body continues to evolve and change in ways that reflect both an internal transformative dynamics (often the focus of psychoanalytic work) and the effect of external processes (most often invoked in social constructionist approaches).

The second proposition, broadly consistent with (if not implicitly contained in) the first, is that the body is not a closed and sealed entity, but a relational 'thing' that is created, bounded, sustained, and ultimately dissolved in a spatiotemporal flux of multiple processes. This entails a relational-dialectical view in which the body (construed as a thing-like entity) internalizes the effects of the processes that create, support, sustain, and dissolve it. The body which we inhabit and which is for us the irreducible measure of all things is not itself irreducible. This makes the body problematic, particularly as 'the measure of all things.'

The body is internally contradictory by virtue of the multiple socio-ecological processes that converge upon it. For example, the metabolic processes that sustain a body entail exchanges with its environment. If the processes change, then the body either transforms and adapts or ceases to exist. Similarly the mix of performative activities available to the body in a given place and time are not independent of the technological, physical, social, and economic environment in which that body has its being. And the representational practices that operate in society likewise shape the body (and in the forms of dress and postures propose all manner of additional symbolic meanings). This means that any challenges to a dominant system of representation of the body (e.g. those mounted by feminists and queer theorists in recent years) become direct challenges to bodily practices. The net effect is to say that different processes (physical and social) 'produce' (both materially and representationally) radically different kinds of bodies. Class, racial, gender, and all manner of other distinctions are marked upon the human body by virtue of the different socio-ecological processes that do their work upon that body.

To put the matter this way is not to view the body as a passive product of external processes. What is remarkable about living entities is the way they capture diffuse energy or information flows and assemble them into complex but well-ordered forms. Creating order out of chaos is, as Prigogyne and Stengers (1984) point out, a vital property of biological systems. As a 'desiring machine' capable of creating order not only within itself but also in its environs, the human body is active and transformative in relation to the processes that produce, sustain, and dissolve it. Thus, bodily persons endowed with semiotic capacities and moral will make their bodies foundational elements in what we have long called 'the body politic.'

To conceptualize the body (the individual and the self) as porous in relation to the environment frames 'self-other' relations (including the relation to 'nature') in a particular way. If, for example, we understand the body to internalize all there is (a strong doctrine of internal relations of the sort I have outlined elsewhere – see Harvey, 1996, Chapter 2) then the reverse proposition also holds. If the self internalizes all things then the self can be 'the measure of all things.' This idea goes back to Protagoras and the Greeks. It allows the individual to be viewed as some kind of decentered center of the cosmos, or, as Munn (1985, 14, 17), in her insightful analysis of social practices on the Melanesian island of Gawa, prefers to put it, 'bodily spacetime serves as a condensed sign of the wider spacetime of which it is a part.' It is only if the body is viewed as being open and porous to the world that it can meaningfully be considered in this way. It is not how the body is seen in the dominant Western tradition. Strathern (1988, 135) underlines the problem:

> The socialized, internally controlled Western person must emerge as a *microcosm of the domesticating process* by which natural resources are put to cultural use ... The only internal relation here is the way a person's parts 'belong' to him or herself. Other relationships bear in from outside. A person's attributes are thus modified by external pressure, as are the attributes of things, but they remain intrinsic to his or her identity.

But in the Melanesian case:

> [The] person is a living commemoration of the actions which produced it ...
> persons are the objectified form of relationships, and it is not survival of the
> self that is at issue but the survival or termination of relations. Eating does
> not necessarily imply nurture; it is not an intrinsically beneficiary act, as it is
> taken to be in the Western commodity view that regards the self as thereby
> perpetuating its own existence. Rather, eating exposes the Melanesian
> person to all the hazards of the relationships of which he/she is com-
> posed ... Growth in social terms is not a reflex of nourishment; rather, in
> being a proper receptacle for nourishment, the nourished person bears
> witness to the effectiveness of a relationship with the mother, father, sister's
> husband or whoever is doing the feeding ... Consumption is not a simple
> matter of self-replacement, then, but the recognition and monitoring of
> relationships ... The self as individual subject exists ... in his or her
> capacity to transform relations.   (Strathern, 1988, 302)

This relational conception of the body, of self, individual, and, consequently, of political identity is captured in the Western tradition only in
dialectical modes of argumentation. Traces of it can also be found in the
contemporary work of deep ecologists (cf. Naess and Rothenberg, 1989)
and the view is now widespread in literary and feminist theory. It constitutes a rejection of the world view traditionally ascribed to Descartes,
Newton, and Locke, which grounds the ideal of the 'civilized' and
'individualized' body (construed as an entity in absolute space and time
and as a site of inalienable and bounded property rights) in much of
Western thought.

It then follows that the manner of production of spacetime is inextricably connected with the production of the body. 'With the advent of
Cartesian logic,' Lefebvre (1991, 1) complains, 'space had entered the
realm of the absolute ... space came to dominate, by containing them, all
senses and all bodies.' Lefebvre and Foucault (particularly in *Discipline
and Punish*) here make common cause: the liberation of the senses and the
human body from the absolutism of that produced world of Newtonian/
Cartesian space and time becomes central to their emancipatory strategies.
And that means challenging the mechanistic and absolute view by means
of which the body is contained and disciplined. But by what bodily
practices was this Cartesian/Newtonian conception of spacetime produced? And how can such conceptions be subverted?

We here encounter a peculiar conundrum. On the one hand, to return to
the human body as the fount of all experience (including that of space and
time) is presently regarded as a means (now increasingly privileged) to
challenge the whole network of abstractions (scientific, social, political-
economic) through which social relations, power relations, institutions,

and material practices get defined, represented, and regulated. But on the other hand, no human body is outside of social processes of determination. To return to it is, therefore, to instantiate the social processes being purportedly rebelled against. If, for example, workers are transformed, as Marx suggests in *Capital*, into appendages of capital in both the work place and the consumption sphere (or, as Foucault prefers it, bodies are made over into *docile bodies* by the rise of a powerful disciplinary apparatus, from the eighteenth century onwards) then how can their bodies be a measure, sign, or receiver of anything outside of the circulation of capital or of the various mechanisms that discipline them? Or, to take a more contemporary version of the same argument, if we are all now *cyborgs* (as Haraway in her celebrated manifesto on the topic suggests), then how can we measure anything outside of that deadly embrace of the machine as extension of our own body and body as extension of the machine?

So while return to the body as the site of a more authentic (epistemological and ontological) grounding of the theoretical abstractions that have for too long ruled purely as abstractions may be justified, that return cannot in and of itself guarantee anything except the production of a narcissistic self-referentiality. Haraway (1991, 190) sees the difficulty. 'Objectivity,' she declares, 'turns out to be about particular and specific embodiment and definitely not about the false vision promising transcendence of all limits and responsibility.' So whose body is it that is to be the measure of all things? Exactly how and what is it in a position to measure? These are deep questions to which we will perforce return again and again. We cannot begin to answer them, however, without some prior understanding of how bodies are socially produced.

2 The theory of the bodily subject in Marx

Let us suppose that Marx's categories are not dismissed as 'thoroughly destabilised.' I do not defend that supposition, though I note that from the *Economic and Philosophical Manuscripts* onwards Marx (1964 edition, 143) grounded his ontological and epistemological arguments on real sensual bodily interaction with the world:

> *Sense-perception* must be the basis of all science. Only when it proceeds from sense-perception in the two-fold form of *sensuous* consciousness and of *sensuous* need – that is, only when science proceeds from nature – is it *true* science.

Marx also elaborated a philosophy of internal relations and of dialectics consistent with the relational conception of the body outlined above (particularly by Strathern). The contemporary rush to return to the body as the irreducible basis of all argument is, therefore, a rush to return to the point where Marx, among many others, began.

While he does not tell us everything we might want to know, Marx does propose a theory of the production of the bodily subject under capitalism. Since we all live within the world of capital circulation and accumulation this has to be a part of any argument about the nature of the contemporary body. To evade it (on the specious grounds that Marx's categories are destabilized or, worse still, outmoded and unfashionable) is to evade a vital aspect of how the body must be problematized. And while Marx's theorizing in *Capital* is often read (incorrectly, as I shall hope to show) as a pessimistic account of how bodies, construed as passive entities occupying particular performative economic roles, are shaped by the external forces of capital circulation and accumulation, it is precisely this analysis that informs his other accounts of how transformative processes of human resistance, desire for reform, rebellion, and revolution can and do occur.

A preparatory step is to broaden somewhat the conventional Marxian definition of 'class' (or, more exactly, of 'class relation') under capitalism to mean *positionality in relation to capital circulation and accumulation*. Marx often fixed this relation in terms of property rights over the means of production (including, in the laborer's case, property rights to his or her own body), but I want to argue that this definition is too narrow to capture the content even of Marx's own analyses (Marx, recall, avoided any formal sociological definitions of class throughout his works). Armed with such a definition of 'positionality with respect to capital circulation and accumulation' we can better articulate the internal contradictions of multiple positionalities within which human beings operate. The laborer as person is a worker, consumer, saver, lover, and bearer of culture, and can even be an occasional employer and landed proprietor, whereas the laborer as an economic role – the category Marx analyses in *Capital* – is singular.

Consider, now, one distinctive systemic concept that Marx proposed. *Variable capital* refers to the sale/purchase and use of labor power as a commodity. But as Marx's analysis proceeds it becomes evident that there is a distinct circulation process to variable capital itself. The laborer (a person) sells labor power (a commodity) to the capitalist to use in the labor process in return for a money wage which permits the laborer to purchase capitalist-produced commodities in order to live in order to return to work ... Marx's distinction between the laborer (*qua* person, body, will) and labor power (that which is extracted from the body of the laborer as a commodity) immediately provides an opening for radical critique. Laborers are necessarily alienated because their creative capacities are appropriated as the commodity labor power by capitalists. But we can broaden the question: what effect does the circulation of variable capital (the extraction of labor power and surplus value) have on the bodies (persons and subjectivities) of those through whom it circulates? The answer initially breaks down into a consideration of what happens at different moments of productive consumption, exchange, and individual consumption.

Productive consumption

Productive consumption of the commodity labor power in the labor process under the control of the capitalist requires, *inter alia*, the mobilization of 'animal spirits,' sexual drives, affective feelings, and creative powers of labor to a given purpose defined by capital. It means: harnessing basic human powers of cooperation/collaboration; the skilling, deskilling, and reskilling of the powers of labor in accord with technological requirements; acculturation to routinization of tasks; enclosure within strict spatiotemporal rhythms of regulated (and sometimes spatially confined) activities; frequent subordinations of bodily rhythms and desires 'as an appendage of the machine;' socialization into long hours of concentrated labor at variable but often increasing intensity; development of divisions of labor of different qualities (depending upon the heterogeneity or homogeneity of tasks, the organization of detailed versus social divisions of labor); responsiveness to hierarchy and submission to authority structures within the work place; separations between mental and manual operations and powers; and, last but not least, the production of variability, fluidity, and flexibility of labor powers able to respond to those rapid revolutions in production processes so typical of capitalist development.

I supply this list (drawn from Marx's *Capital*) mainly to demonstrate how the exigencies of capitalist production push the limits of the working body – its capacities and possibilities – in a variety of different and often fundamentally contradictory directions. On the one hand capital requires educated and flexible laborers, but on the other hand it refuses the idea that laborers should think for themselves. While education of the laborer appears important it cannot be the kind of education that permits free thinking. Capital requires certain kinds of skills but abhors any kind of monopolizable skill. While a 'trained gorilla' may suffice for some tasks, for others creative, responsible workers are called for. While subservience and respect for authority (sometimes amounting to abject submission) is paramount, the creative passions, spontaneous responses, and animal spirits necessary to the 'form-giving fire' of the labor process must also be liberated and mobilized. Healthy bodies may be needed but deformities, pathologies, sickness are often produced. Marx highlights such contradictions:

[L]arge scale industry, by its very nature, necessitates variation of labour, fluidity of functions, and mobility of the worker in all directions. But on the other hand, in its capitalist form it reproduces the old division of labour with ossified particularities. We have seen how this absolute contradiction does away with all repose, all fixity and all security as far as the worker's life situation is concerned ... But if, at present, variation of labour imposes itself after the manner of an overpowering natural law, and with the blindly destructive action of a natural law that meets with obstacles everywhere, large scale industry, through its very catastrophes, makes the recognition of variation of labour and hence of the fitness of the worker for the maximum number of different kinds of labour into a question of life and death.

(Marx, 1976 edition, 617)

Marx sees these contradictions being worked out historically and dialectically (largely though not solely through the use of coercive force and active struggle). But part of what the creative history of capitalism has been about is discovering new ways (and potentialities) in which the human body can be put to use as the bearer of the capacity to labor. Marx observes (1976 edition, 617), for example, that 'technology discovered the few grand fundamental forms of motion which, despite all the diversity of the instruments used, apply necessarily to every productive action of the human body.' Older capacities of the human body are reinvented, new capacities revealed. The development of capitalist production entails a radical transformation in what the working body is about. The unfinished project of the human body is pushed in a particular set of contradictory directions. And a whole host of sciences for engineering and exploring the limits of the human body as a productive machine, as a fluid organism, has been established to explore these possibilities. Gramsci (1971 edition), among others, thus emphasizes again and again how capitalism is precisely about the production of a new kind of laboring body.

While such contradictions may be internalized within the labor force as a whole, this does not necessarily mean that they are internalized within the body of each laborer. Indeed, it is the main thrust of Marx's own presentation that the 'collective body' of the labor force is broken down into hierarchies of skill, of authority, of mental and manual functions, etc. in such a way to render the category of variable capital internally heterogeneous. And this heterogeneity is unstable. The perpetual shifting that occurs within the capitalist mode of production ensures that requirements, definitions of skill, systems of authority, divisions of labor, etc. are never stabilized for long. So while the collective laborer will be fragmented and segmented, the definitions of and relations between the segments will be unstable and the movements of individual laborers within and between segments correspondingly complex. It is not hard to see that in the face of these contradictions and multiple instabilities, capitalism will require some sort of disciplinary apparatus of surveillance, punishment and ideological control that Marx frequently alludes to and which Foucault elaborates upon in ways that I find broadly complementary rather than antagonistic to Marx's project. But the instability never goes away (as witnessed by the whole historical geography of skilling, de-skilling, reskilling, etc.). While the instability is disconcerting, sometimes destructive, and always difficult to cope with, it provides multiple opportunities for subversion and opposition on the part of the laborers.

But whose body is inserted into the circulation of variable capital and with what effects? Marx does not provide any systematic answer to that question in part because this was not the primary object of his theoretical enquiry (he largely dealt with economic roles rather than with persons). Who exactly gets inserted where is a detailed historical-geographical question that defies any simple theoretical answer. Marx is plainly aware that bodies are differentiated and marked by different physical productive capacities and qualities according to history, geography, culture, and tradition. He is also aware that signs of race, ethnicity, age, and gender are used as external measures of what a certain kind of laborer is capable of or permitted to do. The incorporation of women and children into the circulation of variable capital in nineteenth-century Britain occurred for certain distinctive reasons that Marx is at pains to elaborate upon. This in turn provoked distinctive effects, one of which was to turn the struggle over the length of the working day and the regulation of factory employment into a distinctive struggle to protect women and children from the impacts of capitalism's 'werewolf hunger' for surplus value. The employment of women and children as wage laborers, furthermore, not only provided 'a new foundation for the division of labor' (Marx, 1976, 615), it also posed (and continues to pose) a fundamental challenge to many traditional conceptions of the family and of gender roles:

> However terrible and disgusting the dissolution of the old family ties within the capitalist system may appear, large scale industry, by assigning an important part in socially organized processes of production, outside the sphere of the domestic economy, to women, young persons and children of both sexes, does nevertheless create a new economic foundation for a higher form of the family and of relations between the sexes . . . It is also obvious that the fact that the collective working group is composed of individuals of both sexes and all ages must under the appropriate conditions turn into a source of humane development, although in its spontaneously developed, brutal, capitalist form, the system works in the opposite direction, and becomes a pestiferous source of corruption and slavery, since here the worker exists for the process of production, and not the process of production for the worker.

In remarks on slavery, colonialism, and immigrants (e.g. the Irish into Britain), Marx makes clear that constructions of race and ethnicity are likewise implicated in the circulation process of variable capital. Insofar as gender, race, and ethnicity are all understood as social constructions rather than as essentialist categories, so the effect of their insertion into the circulation of variable capital (including positioning within the internal heterogeneity of collective labor and, hence, within the division of labor and the class system) has to be seen as a powerful force reconstructing them in distinctively capitalist ways.

There are a number of corollaries. Firstly, the productiveness of a person gets reduced to the ability to produce surplus value. To be a productive worker, Marx (1976, 644) ironically notes, 'is therefore not a piece of luck but a misfortune;' the only value that the laborer can have is not determined in terms of work done and useful social effect but through 'a specifically social relation of production ... which stamps the worker as capital's direct means of valorization.' The gap between what the laborer as person might desire and what is demanded of the commodity labor power extracted from his or her body is the nexus of alienation. And while workers as persons may value themselves in a variety of ways depending upon how they understand their productivity, usefulness and value to others, the more restricted social valuation given by their capacity to produce surplus value for capital necessarily remains central to their lives (as even highly educated middle-level managers find out when they, too, are laid off). Exactly what that value is, however, depends on conditions external to the labor process, hinging, therefore, upon the question of exchange.

Secondly, lack of productivity, sickness (or of any kind of pathology) gets defined within this circulation process as inability to go to work, inability to perform adequately within the circulation of variable capital (to produce surplus value) or to abide by its disciplinary rules (the institutional effects elaborated on by Rothman [1971] and Foucault [1995] in the construction of asylums and prisons are already strongly registered in Marx's chapters on 'The Working Day' and the 'So-Called Primitive Accumulation'). Those who cannot (for physical, psychic, or social reasons) continue to function as variable capital, furthermore, fall either into the 'hospital' of the industrial reserve army (sickness is defined under capitalism broadly as inability to work) or else into that undisciplined inferno of the lumpenproletariat (read 'underclass') for whom Marx regrettably had so little sympathy. The circulation of variable capital, being so central to how capitalism operates as a social system, defines roles of employed 'insiders' and unemployed 'outsiders' (often victimized and stigmatized) that have ramifications for society as a whole. This brings us back to the moment of 'exchange.'

Exchange of variable capital

The commodity which the laborer (*qua* person) exchanges with the capitalist is labor power, the capacity to engage in concrete labor. The basic condition of the contract is supposedly that the capitalist has the right to whatever the laborer produces, has the right to direct the work, determine the labor process, and have free use of the capacity to labor during the hours and at the rate of remuneration stipulated in the contract. The rights of capital are frequently contested and it is interesting to see on what grounds. While capitalists may have full rights to the commodity labor power, they do not have legal rights over the person of the laborer (that would be slavery). Marx insists again and again that this is a fundamental principle of wage labor under capitalism.

The laborer as person should have full rights over his or her own body and should always enter the labor market under conditions of freedom of contract even if, as Marx (1976, 272–3) notes, a worker is 'free in the double sense that as a free individual he can dispose of his labour-power as his own commodity, and that, on the other hand, he has no other commodity for sale, i.e. he is rid of them, he is free of all the objects needed for the realization of his labour power.' But the distinction between laborer as person and labor power has further implications. The capitalist has not the formal right to put the body of the person at risk, for example, and working practices that do so are open to challenge. This principle carries over even into the realm of the cultural and bodily capital (as Bourdieu defines them): hence much of the resistance to de-skilling, redefinitions of skill, etc. Of course, these legalities are continually violated under capitalism and situations frequently do arise in which the body and person of the laborer is taken over under conditions akin to slavery. But Marx's point is that preservation of the integrity and fullness of the laboring person and body within the circulation process of variable capital is the fulcrum upon which contestation and class struggle both within and without the labor process occurs. Even bourgeois legality (as incorporated in the Factory Acts then and in, say, Occupational Safety and Health regulations now) has to concede the difference between the right to the commodity labor power and the non-right to the person who is bearer of that commodity.

This struggle carries over into the determination of the value of variable capital itself, because here the 'neediness' of the body of the laborer forms the datum upon which conditions of contract depend. In *Capital*, Marx, for purposes of analysis, presumes that in a given place and time such needs are fixed and known (only in this way can he get a clear fix upon how capital is produced through surplus value extraction). But Marx well understood that these conditions are never fixed but depend on physical circumstances (e.g. climate), cultural and social conditions, the long history of class struggle over what is a liveable wage for the laborer, as well as upon a moral conception as to what is or is not tolerable in a civilized society. Consider how Marx (1976, 341) presents the matter in his chapter on 'The Working Day':

> During part of the day the vital force must rest, sleep; during another part the man has to satisfy physical needs, to feed, wash and clothe himself. Besides these purely physical limitations, the extension of the working day encounters moral obstacles. The worker needs time in which to satisfy his intellectual and social requirements, and the extent and number of those requirements is conditioned by the general level of civilization. The length of the working day therefore fluctuates within boundaries that are physical and social.

Marx's primary point of critique of capitalism is that it so frequently violates, disfigures, subdues, maims, and destroys the integrity of the laboring body (even in ways that can be dangerous to the further accumulation of capital). It is, furthermore, in terms of the potentialities and possibilities of that laboring body (its 'species being' as Marx [1964 edition] called it in his early work) that the search for an alternative mode of production is initially cast.

But surplus value depends upon the difference between what labor gets (the value of labor power) and what labor creates (the value of the commodity produced). The use value of the commodity labor power to the capitalist is that it can engage in concrete labor in such a way as to embed a given amount of abstract labor in the commodity produced. For the capitalist it is abstract labor that counts and the value of labor power and the concrete practices of the laborer are disciplined and regulated within the circulation of variable capital by the 'laws of value' which take abstract labor as their datum.

Abstract labor – value – is measured through exchange of commodities over space and time and ultimately on the world market. Value is a distinctive spatiotemporal construction depending upon the development of a whole array of spatiotemporal practices (including the territorialization of the earth's surface through property rights and state formation and the development of geographical networks and systems of exchange for money and all commodities, including that of labor power itself). The value of labor power to the capitalist is itself contingent upon the realization of values across a world of socially constructed spatiotemporal political-economic practices. This limits the value that the laborer can acquire in a particular place both in production and in the market. Furthermore, the conditions of exchange of labor power are limited in labor markets both by

systematic biases (gender and racial disparities in remuneration for comparative work are well documented) and by mobilization of an industrial
reserve army (either *in situ* or through the migratory movements of both
capital and labor searching for 'better' contractual conditions).

It is exactly at this point that the connection between what we now refer
to as 'globalization' (see Chapter 4) and the body becomes explicit. But
how should this be thought about? Marx depicts the circulation of variable
capital as a 'commodity for commodity' exchange: the worker exchanges
the use value of labor power for the use value of the commodities that can
be bought for the money wage. Exchanges of this sort are usually highly
localized and place-specific. The worker must take his or her body to work
each day (even under conditions of telecommuting). But labor power is
inserted as a commodity into a Money-Commodity-Money circulation
process which easily escapes the spatiotemporal restraints of local labor
markets and which makes for capital accumulation on the world stage.
Accumulation accelerates turnover time (it shortens working periods,
circulation times, etc.) while simultaneously annihilating space through
time while preserving certain territorialities (of the factory and the nation
state) as domains of surveillance and social control. Spatiotemporality
defined at one scale (that of 'globalization' and all its associated meanings)
intersects with bodies that function at a much more localized scale.
Translation across spatiotemporal scales is here accomplished by the inter-
section of two qualitatively different circulation processes, one of which is
defined through the long historical geography of capital accumulation
while the other depends upon the production and reproduction of the
laboring body in a far more restricted space. This leads to some serious
disjunctions, of the sort that Hareven (1982) identifies in her analysis of
Family Time and Industrial Time. But as Hareven goes on to show, these
two spatiotemporal systems, though qualitatively different from each
other, have to be made 'cogredient' or 'compossible' (see Harvey, 1996,
for a fuller explication of these terms) with each other. Thus do links
between the 'local' and the 'global' become established. Different bodily
qualities and modes of valuation (including the degree of respect for the
bodily integrity and dignity of the laborer) achieved in different places are
brought into a spatially competitive environment through the circulation
of capital. Uneven geographical development of the bodily practices and
sensibilities of those who sell their labor power becomes one of the defining
features of class struggle as waged by both capital and labor.

Put in more direct contemporary terms, the creation of unemployment
through down-sizing, the redefinitions of skills and remunerations for
skills, the intensification of labor processes and of autocratic systems of
surveillance, the increasing despotism of orchestrated detailed divisions of

labor, the insertion of immigrants (or, what amounts to the same thing, the migration of capital to alternative labor sources), and the coerced competitive struggle between different bodily practices and modes of valuation achieved under different historical and cultural conditions, all contribute to the uneven geographical valuation of laborers as persons. The manifest effects upon the bodies of laborers who live lives embedded in the circulation of variable capital is powerful indeed. Sweatshops in New York mimic similar establishments in Guatemala and subject the workers incorporated therein to a totalizing and violently repressive regime of body disciplines. The construction of specific spatiotemporal relations through the circulation of capital likewise constructs a connection between the designer shirts we wear upon our backs, the Nike shoes we sport, and the oriental carpets upon which we walk, and the grossly exploited labor of tens of thousands of women and children in Central America, Indonesia, and Pakistan (just to name a few of the points of production of such commodities).

The moment of consumption

The laborer does not only lie in the path of variable capital as producer and exchanger. He/she also lies in that circulation process as consumer and reproducer of self (both individually and socially). Once possessed of money the laborer is endowed with all the autonomy that attaches to any market practice:

> It is the worker himself who converts the money into whatever use-values he desires; it is he who buys commodities as he wishes and, as the *owner of money*, as the buyer of goods, he stands in precisely the same relationship to the sellers of goods as any other buyer. Of course, the conditions of his existence – and the limited amount of money he can earn – compel him to make his purchases from a fairly restricted selection of goods. But some variation is possible as we can see from the fact that newspapers, for example, form part of the essential purchases of the urban English worker. He can save and hoard a little. Or else he can squander his money on drink. Even so, he acts as a free agent; he must pay his own way; he is responsible to himself for the way he spends his wages.   (Marx, 1976, 1,033)

This is an example of Marx's tacit appeal to 'positionality in relationship to capital accumulation' as a practical definition of class relations. As the focus shifts so does the meaning of class positionality. The laborer has limited freedom to choose not only a personal lifestyle but also, through the collective exercise of demand preferences, he/she can express his/her desires (individually and collectively) and thereby influence the capitalist choice of what to produce. Elaboration on that idea permits us to see, as we look at the circulation of variable capital as a whole, that what is true for the individual laborer is rather more limited when looked at from the standpoint of the collectivity:

> The capitalist class is constantly giving to the working class drafts, in the form of money, on a portion of the product produced by the latter and appropriated by the former. The workers give these drafts back just as constantly to the capitalists, and thereby withdraw from the latter their allotted share of their own product ... The individual consumption of the worker, whether it occurs inside or outside the workshop, inside or outside the labour process, remains an aspect of the production and reproduction of capital ... From the standpoint of society, then, the working class, even when it stands outside the direct labour process, is just as much an appendage of capital as the lifeless instruments of labour are.
>
> (Marx, 1976, 713, 719)

Deeper consideration of what amounts to a 'company store' relation between capital and labor is instructive. The disposable income of the laborers forms an important mass of effective demand for capitalist output (this is the relation that Marx explores at great length in Volume 2 of *Capital*). Accumulation for accumulation's sake points towards either an increasing mass of laborers to whom necessities can be sold or a changing standard of living of the laborers (it usually means both). The production of new needs, the opening up of entirely new product lines that define different lifestyles and consumer habits, is introduced as an important means of crisis avoidance and crisis resolution. We can then see more clearly how it is that variable capital has to be construed as a circulation process (rather than as a single causal arrow) for it is through the payment of wages that the disposable income to buy the product of the capitalists is partially assured.

But all of this presumes 'rational consumption' on the part of the laborer – rational, that is, from the standpoint of capital accumulation (Marx, 1978 edition, 591). The organization, mobilization, and channeling of human desires, the active political engagement with tactics of persuasion, surveillance, and coercion, become part of the consumptuary apparatus of capitalism, in turn producing all manner of pressures on the body as a site of and a performative agent for 'rational consumption' for further accumulation (cf. Henry Ford's obsession with training social workers to monitor the budgets of his workers).

But the terms of 'rational consumption' are by no means fixed, in part because of the inevitable destabilizing effects of perpetual revolutions in capitalist technologies and products (revolutions which affect the household economy as well as the factory), but also because, given the discretionary element in the worker's use of disposable income, there is as much potential for social struggle over lifestyle and associated bodily practices as there is in the realm of production itself. Struggles over the social wage – over, for example, the extent, direction, and distributional effects of state expenditures – have become critical in establishing the baseline of what might be meant by a proper standard of living in a 'civilized' country. Struggles over the relation between 'housework' and 'labor in the market' and the gender allocation of tasks within domestic settings also enter into the picture (cf. Marx's 1976 edition, 518, commentary on how the importance of domestic labor gets 'concealed by official political economy' and the revived debate in the 1970s on the role of housework in relation to the circulation of variable capital).

This moment in the circulation of variable capital, though not totally absent in Marx's account, is not strongly emphasized. With the United States (and, presumably, much of the advanced capitalist world) in mind, Lowe (1995, 67) now argues that:

> Lifestyle is the social relations of consumption in late capitalism, as distinct from class as the social relations of production. The visual construction and presentation of self in terms of consumption relations has by now over-shadowed the class relations of production in the workplace ... [Consumption] is itself dynamically developed by the design and production of changing product characteristics, the juxtaposition of image and sign in lifestyle and format, and the segmentation of consumer markets.

This suggests a double contradiction within the advanced capitalist world (and a nascent contradiction within developing countries). First, by submitting unquestioningly and without significant struggle to the dictates of capital in production (or by channeling struggle solely to the end of increasing disposable income), workers may open for themselves wider terrains of differentiating choice (social or individual) with respect to lifestyle, structures of feeling, household organization, reproductive activities, expressions of desire, pursuit of pleasures, etc. within the moment of consumption. This does not automatically deliver greater happiness and satisfaction. As Marx (1965 edition, 33) notes:

> [A]lthough the pleasures of the labourer have increased, the social gratification which they afford has fallen in comparison with the increased pleasures of the capitalist. Our wants and pleasures have their origin in society; we therefore measure them in relation to society; we do not measure them in relation to the objects which serve for their gratification. Since they are of a social nature, they are of a relative nature.

Conversely, by locking workers into certain conceptions of lifestyle, consumer habits, and desire, capitalists can more easily secure compliance within the labor process while capturing distinctive and proliferating market niches for their sales.

Struggles arise between how workers individually or collectively exercise their consumer and lifestyle choices and how capitalist forces try to capture and guide those choices towards rational consumption for sustained accumulation. Marx does not scrutinize such conflicts but no particular difficulty attaches to integrating them into his framework. Plainly, the process is marked by extraordinary heterogeneity at the same time as it is fraught with instability. For example, whole communities of lifestyle (such as those shaped by working classes in industrial settings or by distinctive cultural traditions) may be created within the circulation of variable capital only ultimately to be dissolved (even in the face of considerable resistance) by the same processes that led to their initial formation. The recent history of deindustrialization is full of examples of this.

A wide range of bodily practices and cultural choices with respect to consumption can in principle be embedded in the circulation of variable capital. The range depends, of course, upon the amount of discretionary income in the laborer's possession (and, plainly, the billion or so workers living on less than a dollar a day cannot exercise anywhere near the amount of influence as well-paid workers in the advanced capitalist countries). Variable capital does not determine the specific nature of consumer choices or even of consumer culture, though it certainly works to powerful effect. This means that production must internalize powerful effects of heterogeneous cultural traditions and consumer choices, whether registered collectively through political action (to establish a 'social wage' through welfare programs) or individually through personal consumption choices. It is in this sense that it is meaningful to speak of the moments of production and consumption as a matter of internal relations, the one with the other.

The circulation of variable capital as a whole
Consider, then, the figure of the laborer caught within the rules of circulation of variable capital as a whole. The experiential world, the physical presence, the subjectivity and the consciousness of that person are partially if not predominantly forged in the fiery crucible of the labor process, the passionate pursuit of values and competitive advantage in labor markets, and in the perpetual desires and glittery frustrations of commodity culture. They are also forged in the matrix of time-space relations between persons largely hidden behind the exchange and movement of things. The evident instabilities within the circulation of variable capital coupled with the different windows on the world constructed through moments of production, exchange, and consumption place the

laboring body largely at the mercy of a whole series of forces outside of any one individual's control. It is in this sense that the laboring body must be seen as an internal relation of the historically and geographically achieved processes of capital circulation.

When, however, we consider the accumulation process as a whole, we also see that 'the maintenance and reproduction of the working class remains a necessary condition for the reproduction of capital.' The working class is, in effect, held captive within a 'company store' relation to capital accumulation that renders it an appendage of capital at all moments of its existence. The capitalist, in short, 'produces the worker as wage laborer.' Marx (1973 edition, 717–18) continues:

> The capital given in return for labour-power is converted into means of subsistence which have to be consumed to reproduce the muscles, nerves, bones and brains of existing workers, and to bring new workers into existence. Within the limits of what is absolutely necessary, therefore, the individual consumption of the working class is the reconversion of the means of subsistence given by capital in return for labour-power into fresh labour power which capital is then again able to exploit. It is the production and reproduction of the capitalist's most indispensable means of production: the worker.

The issue of reproduction is then immediately posed. Marx was less than forthcoming on this question leaving it, as the capitalist does, 'to the worker's drives for self preservation and propagation.' The only rule he proposes is that the laboring family, denied access to the means of production, would strive in times of prosperity as in depression, to accumulate the only form of 'property' it possessed: labor power itself. Hence arises a connexion between expanded accumulation and 'maximum growth of population – of living labor capacities' (Marx, 1973 edition, 608).

But it is also clear that as laborers acquire property on their own account or move to acquire cultural as well as 'human capital' in the form of skills, that this equation will likely change and generate different reproductive strategies, together with different objectives for social provision through class struggle within the working classes of the world. Furthermore, Marx's occasional commentaries on 'the family' as a socially constructed unit of reproduction (coupled with Engels's treatise on *Origin of the Family, Private Property and the State* with its emphasis upon division of labor between the sexes and propagation of the species) indicates a material point at which questions of sexuality and gendering intersect with political economy. Elaborations by socialist feminists in recent years here assume great importance. If the circulation of variable capital as a whole is about the reproduction of the working class in general, then the question of the conditions of its biological and social reproduction must be posed in ways that acknowledge such complexities (cf. the controversy between Butler, 1998, and Fraser, 1997).

Potentialities for reaction and revolt against capital get defined from the different perspectives of production, exchange, consumption, or reproduction. Nevertheless, in aggregate we can still see how the pernicious capitalistic rules that regulate the process of circulation of variable capital as a whole operate as a constructive/destructive force (both materially and representationally) on laboring bodies across these different moments. Capital continuously strives to shape bodies to its own requirements, while at the same time internalizing within its *modus operandi* effects of shifting and endlessly open bodily desires, wants, needs, and social relations (sometimes overtly expressed as collective class, community, or identity-based struggles) on the part of the laborer. This process frames many facets of social life, such as 'choices' about sexuality and biological reproduction or of culture and ways of life even as those 'choices' (if such they really are) get more generally framed by the social order and its predominant legal, social, and political codes, and disciplinary practices (including those that regulate sexuality).

Study of the circulation of variable capital cannot, in and of itself, tell us everything we need to know. It is, to begin with, just one subset of a slew of different circulation processes that make up the circulation of capital in general. Productive, finance, landed, and merchant capitals all have their own modalities of motion and the circulation of bourgeois revenues generates complex relations between 'needs,' 'wants,' and 'luxuries' that affect lifestyle choices, status symbols, and fashions as set by the rich, powerful, and famous. These set relative standards for the laboring poor since, as Marx also insists, the sense of well-being is a comparative rather than an absolute measure and the gap between rich and poor is just as important as the absolute conditions of sustenance. Furthermore, the mediating activities of states (as registered through the circulation of tax revenues and state-backed debt) in determining social wages and setting 'civilized' and 'morally acceptable' standards of education, health, housing, etc. play crucial roles on the world stage of capital accumulation and in setting conditions within which the circulation of variable capital can occur. The point here is not to insist on any complete or rigorous accounting – either theoretical or historical – of these intersecting processes. But an understanding of the conditions of circulation of variable capital is indisputably a necessary condition for understanding what happens to bodies in contemporary society.

There are innumerable elaborations, modifications, reformulations, and even outright challenges to Marx's limited but tightly argued theory of the production of the laboring body and of individual and collective subjectivities. There is much that is lacking (or only lightly touched upon) in Marx's schema, including the sexual and erotic, the gendering and racial identifications of bodies, the psychoanalytic and representational, the linguistic and the rhetorical, the imaginary and the mythical (to name just a few of the obvious absences). The roles of gender within the spatial and social divisions of labor have been the focus, for example, of a considerable range of studies in recent years (see, e.g., Hanson and Pratt, 1994) and the question of race relations or ethnic/religious discriminations within segmented labor markets has likewise been brought under the microscope (see, e.g., Goldberg, 1993) in ways that have given much greater depth and purpose to Marx's (1976 edition, 414) observation that 'labour in a white skin cannot emancipate itself where it is branded in a black skin.' So there are plenty of other processes – metabolic, ecological, political, social, and psychological – that play key roles in relation to bodily practices and possibilities.

But these absences cannot be cured by an erasure of either the method or substance of Marx's approach. The latter is something to build upon rather than to negate. The human body is a battleground within which and around which conflicting socio-ecological forces of valuation and representation are perpetually at play. Marx provides a rich conceptual apparatus to understand processes of bodily production and agency under capitalism. Just as important, he provides an appropriate epistemology (historical-geographical as well as dialectical) to approach the question of how bodies get produced, how they become the signifiers and referents of meanings, and how internalized bodily practices might in turn modify the processes of their self-production under contemporary conditions of capitalistic globalization.

CHAPTER 7

Body politics and the struggle for a living wage

1 The political body in the body politic

Bodies embedded in a social process such as the circulation of variable capital are never to be construed as docile or passive. It is, after all, only through the 'form-giving fire' of the capacity to labor that capital is produced. And even if labor under the domination of capital is condemned for the most part to produce the conditions and instruments of its own domination (as much in the realm of consumption and exchange as in production itself), the transformative and creative capacities of the laborer always carry the potentiality (however unimaginable in the present circumstances) to fashion an alternative mode of production, exchange, and consumption. Those transformative and creative capacities can never be erased. This poses acute problems for the maintenance of capitalism's authority while providing multiple opportunities for laborers to assert their agency and will. It is no accident, therefore, that Marx attaches the appellation 'living' to the labor embedded in the circulation of variable capital to emphasize not only its fundamental qualities of dynamism and creativity but also to indicate where the life-force and the subversive power for change resides.

An analysis of the circulation of variable capital shows that 'body politics' looks different from the standpoints of production, exchange, and consumption. Trade-offs plainly exist between how laborers submit to or struggle with the dictates of capital at one moment to enhance their powers at another. Abject submission to the dictates of capital within production, for example, may for some be a reasonable price to bear for adequate pleasures and fulfillment of desires (presuming such are possible given the multiple fetishisms of the market) in the realm of consumption. But what dictates whether that price is judged too high? The working body is more than just 'meat' as William Gibson so disparagingly refers to it in his dystopian novel *Neuromancer* and laborers are more than just 'hands' (presuming they have neither head nor belly as Charles Dickens mockingly observes in *Hard Times*). The concept of the body is here in danger of losing its political purchase because it cannot provide a basis to define the *direction* as opposed to the *locus* of political action. Those (like Foucault and Butler) who appeal to the body as a foundational concept consequently experience intense difficulty in elaborating a politics that focuses on anything other than sexuality. Concern for the broader issues of what happens to bodies inserted into the circulation of variable capital typically disappears in such accounts (although Butler [1998] has recently taken pains to point out the connections between body politics and political economic questions). Yet a concept of variable capital which posits the laborer as the pure subject of capital accumulation cannot help solve the problem either. 'Body politics' in this narrow reductionist sense then becomes just as disempowering *vis-à-vis* capital accumulation as the idea of globalization. Something else is required to translate from the realm of body as 'meat' for accumulation to the concept of laborer as political agent.

The body cannot be construed as the locus of political action without a notion of what it is that 'individuals,' 'persons,' or social movements might want or be able to do in the world. Concepts such as *person*, *individual, self,* and *identity,* rich with political thought and possibilities, emerge phoenix-like out of the ashes of body reductionism to take their places within the firmament of concepts to guide political action. Marx has this in mind as he contrasts the deadly passivity of the concept of variable capital with the concept of 'living labor' or, more broadly, of 'class for itself' struggling to redefine the historical and geographical conditions of its own embeddedness within capital accumulation. It is the laborer as *person* who is the bearer of the commodity labor power and that person is the bearer of ideals and aspirations concerning, for example, the dignity of labor and the desire to be treated with respect and consideration as a whole living being, and to treat others likewise.

Some may be tempted at this point to abandon the relational view for, as Eagleton (1997, 22) complains, 'to dissolve human beings to nexuses of processes may be useful if you had previously thought of them as solitary atoms, but unhelpful when you want to insist on their moral autonomy.' Marx (1973 edition, 84) demurs:

> [T]he more deeply we go back into history, the more does the individual, and hence also the producing individual, appear as dependent, as belonging to a greater whole ... Only in the eighteenth century, in 'civil society', do the various forms of social connectedness confront the individual as a mere means towards his private purposes, as external necessity. But the epoch that produces this standpoint, that of the isolated individual, is also precisely that of the hitherto most developed social (from this standpoint, general) relations. The human being is in the most literal sense a [political animal], not merely a gregarious animal, but an animal that can individuate

itself only in the midst of society. Production by an outside individual outside society . . . is as much an absurdity as is the development of language without individuals living *together* and talking to each other.

Marx here builds on Aristotle's view that human beings are both social and political animals needing intimate relations with others and that such forms of social relating constitute and sustain civil society. How human beings have gone about this task has varied historically and geographically. The sense of self and of personhood is relational and socially constructed (and Marx here anticipates Strathern's formulation cited above) in exactly the same way as the body is a social construct except that the forces at work (and it is no accident that Marx cites language as his parallel) are significantly different. The notion of 'individuals possessed of moral autonomy,' for example, is not a universal but arose in the eighteenth century in Europe as commodity exchange and capital accumulation became more generalized. The task of active politics, in Marx's view, is to seek transformations of social relations in the full recognition that the starting point of political action rests upon achieved historical-geographical conditions.

We here encounter a reflexive point from which to critique certain versions of that 'return to the body' that has been so strongly evidenced in recent years. The dangers of 'body reductionism' – the idea that the body is the *only* foundational concept we can trust in looking for an alternative politics – become plain to see. But, in contrast, in searching for associative concepts (such as those of 'person', 'self', and 'individual') there is an equal danger of reconstituting the liberal eighteenth-century ideal of the 'individual' endowed with 'moral autonomy' as the basis for political theory and political action. We have to find a path between 'body reductionism' on the one hand and merely falling back into what Benton (1993, 144) calls 'the liberal illusion' about political rights propagated with such devastating effects through the crude association of capitalism and bourgeois democracy on the other:

> In societies governed by deep inequalities of political power, economic wealth, social standing and cultural accomplishments, the promise of equal rights is delusory with the consequence that for the majority, rights are merely abstract, formal entitlements with little or no *de facto* purchase on the realities of social life. In so far as social life is regulated by these abstract principles and in so far as the promise is taken for its fulfillment, then the discourse of rights and justice is an ideology, a form of mystification which has a causal role in binding individuals to the very conditions of dependence and impoverishment from which it purports to offer emancipation.

The need for the relational view does not disappear but deepens. For while Benton has one side of the picture he loses sight of the ways in which socially embedded notions of personal autonomy and of the power of individuals to regulate their own lives in accordance with their own beliefs and desires can also operate as persistent even if subterranean pressures subverting dominant ideologies in surprising ways. Marx (1964 edition, 181) pioneered such a relational conception in his early works when, for example, he argued:

> To say that man is a *corporeal,* living, real, sensuous, objective being full of natural vigor is to say that he has *real, sensuous, objects* as the objects of his being or of his life, or that he can only *express* his life in real, sensuous objects. *To be* objective, natural and sensuous, and at the same time to have object, nature and sense outside oneself, or oneself to be object, nature and sense for a third party, is one and the same thing ... A being which does not have its nature outside itself is not a natural being, and plays no part in the system of nature. A being which has no object outside of itself is not an objective being. A being which is not itself an object for some third being has no being for its *object*; i.e. it is not objectively related. Its being is not objective. An unobjective being is a nullity – an *un-being.*

While the prose is convoluted the meaning is clear enough – no body exists outside of its relations with other bodies and the exercise of powers and counterpowers among bodies is a central constitutive aspect of social life. In more recent times we can see in Ricoeur's (1992) trenchant criticism of Parfitt and, by implication, Locke and Hume, a critical reminder of how the clash between the liberal conception of personal identity and, in Ricoeur's case, a relational conception of narrative identity produces a dramatically alternative reading of how body politics might be constructed.

All of this returns us, though via a different path, to the point at which we arrived in our analysis of the phenomenon of globalization. From the standpoint of the laborer, embedded as a political person within the circulation of capital, politics is rooted in the positionalities that he or she assumes and the potentialities that attach thereto. On the one hand there is the revolutionary urge to become free of that embeddedness within the circulation of capital that so circumscribes life chances, body politics, and socio-ecological futures. On the other, there is the reformist demand for fair and proper treatment within that circulation process, to be free, for example, of the ugly choice between adequate remunerations in consumption and abject submission in production. And for those billion or so workers in the world who must live on less than a dollar a day (cf. Chapter 3), the struggle for dignity in the workplace, for adequate life chances, for a living wage, and for some broader conception of human, civil, and political rights becomes a minimalist political program. But different moments generate different political arguments and so the potential coherency and singularity of the worker's voice has the awkward habit of dissolving into different opinions as political persons choose their positions and assumptions about identities and interests (cf. Unger, 1987b, 548). Such politics, as I argued at the end of Chapter 3, are necessarily a global as well as a local affair. So it is to a local manifestation of such a struggle that I now turn.

2 Struggling for a living wage

Ever since Thomas Hobbes roundly declared that 'the value of a man is his price,' the question of the proper value of labor power has hovered over capitalism as a problem as difficult to resolve theoretically as it has been practically. The classical political economists could never quite resolve the confusion that arose from on the one hand equating value with labor and on the other hand having to recognize that the value of labor as an input to production was somehow less than the value it generated (thus leaving room for rents, profits, interest, and the like). Marx neatly solved that problem by recognizing a difference between labor as the substance of value and labor power (the capacity to create value) as a commodity sold by laborers to capitalists. Equally neatly, the neoclassicals eviscerated the political message that came from Marx's formulation by equating proper wages with the marginal return on labor as an input to production (leaving open therefore the possibility for a 'fair' rate of return for capital and land). That idea never worked well for, as Marx pointed out, labor is not a commodity like any other. A host of moral, social, historical and geographical circumstances enters into its formulation and valuation. Chief among these is a long and widespread historical geography of class struggle.

In the United States, for example, the concept of an adequate 'living wage' (alongside that of a socially regulated working day) was fundamental to the agitation that began in cities like Baltimore and Pittsburgh with the massive railroad strike of 1877. As Glickman (1997) shows, this was the kind of agitation that ultimately led to minimum wage legislation, at first at State and then subsequently at the Federal level during the New Deal years.

There has always been controversy as to what properly constitutes a living wage. Since 1968, as Pollin and Luce (1998) document, the value of the minimum wage established at the Federal level has declined by some thirty percent in real terms, placing those with full time minimum wage jobs now well below the poverty level. Its 1997 increase (to $5.15 from a baseline of $4.25 an hour in 1994) still kept it well below 1968 standards. With a good deal of frustration at the ability to assure an adequate living wage at the Federal scale, a whole series of local campaigns and agitations at a more local level have in recent years broken out across the United States. One of the pioneers in this movement exists in my home town of Baltimore. I provide, then, an account of this local struggle as an illustration of how a theory of uneven geographical developments might work in conjunction with arguments for a universal system of human rights (cf. Chapter 5).

The circumstances regulating wages and living conditions in Baltimore underwent significant alterations from the late 1960s onwards (see Chapter 8). Severe deindustrialization of the economy (connected with processes of globalization) meant some radical shifts in the circulation of variable capital within the metropolitan region. In addition to widespread structural unemployment (and the production of a so-called and much stigmatized 'underclass') the effect was to move employment away from the blue collar (largely white male and unionized) industrial sector and into a wide array of service activities, particularly those connected to the so-called 'hospitality sector' (hotels, tourism, conventions, museums) that underpinned the redevelopment effort in Baltimore. The result (in line with much of the US economy – see, e.g., Wilson, 1996, and Kasarda, 1995) was widespread long-term structural unemployment and a shift towards non-unionized and female employment in low-paying 'unskilled' jobs. Low-income job opportunities arose in areas such as cleaning, janitorial, parking, and security services. Paying only minimum wages and often resting on temporary work which yielded even less on a weekly basis (with no health, security, or pension benefits) the growth of this form of employment produced an increasing number of 'working poor' – individuals or families fully employed whose incomes were often well below the official poverty line (a recent report put the number of children of the working poor in the United States at 5.6 million in 1994 as opposed to 3.4 million in 1974 – see Holmes, 1996). African-American women, drawn from the impoverished zones of the inner city, became the main source of this kind of labor in Baltimore, indicating a discursive and largely racist-sexist construction of the inherent 'value' of *that* kind of labor power from *that* kind of place. This stereotyping was automatically reinforced and framed within a circulation process of variable capital and capital accumulation that insisted that this was the kind of labor power that was essential to its own valorization.

These broad economic trends were paralleled by a nation-wide political attack upon working-class institutions and government supports (see, e.g., Edsall, 1984) and a general shift by a whole range of public and private institutions towards political-economic practices that emphasized capital accumulation. One effect was spiraling social inequalities of the sort symbolized by the declining value of the minimum wage in real value terms.

A particular instance of this political economic shift is worth recording. In 1984, the Johns Hopkins University and the Johns Hopkins Hospital (both non-profit and educational institutions) in Baltimore formed a for-profit wholly-owned subsidiary called Dome Corporation, which provides security, parking, cleaning, and janitorial services through another subsidiary called Broadway Services Inc. This firm does some of the cleaning and janitorial work in the Johns Hopkins System as well as in a number of City schools, downtown offices, and the like. Most of the employees are women and African-American, drawn from the impoverished zones of Baltimore City. Most were paid at or slightly above the then-prevailing minimum wage of $4.25 (raised to $4.75 in 1996 and then $5.15 in 1997). Full-time employees paid circa $5 per week for minimal health insurance, but a significant portion of the work was done by temporary workers with no benefits. The Johns Hopkins System has by this strategy achieved cost-savings on its cleaning bills and a healthy rate of return (circa 10%) on its investment (debt plus equity). It has since been cited by other universities as a successful model of how to cut costs by out-sourcing its cleaning work while also making a profit.

This is an example of how shifts in the circulation of variable capital can occur. Such shifts have radical effects upon bodily conditions and practices. Everyone recognizes that $4.75 an hour is insufficient to live on. To bring a family of four above the official poverty line would require a permanent job at a mimimum of $7.70 per hour (1996 values) plus benefits, in Baltimore. The lack of health benefits and elementary care translates into a chronic epidemiological condition for many inner-city neighborhoods (and the sad paradox of cleaners unable to use the services of the hospital they clean). The need to hold down two jobs to survive translates into a condition of permanent physical exhaustion from a twelve-hour working day plus travel time on unreliable public transport between job sites and residences. When two jobs could not be had, the effect was to force some of the employed to live in shelters rather than regular housing and eat at charity soup kitchens rather than at Roy Rogers or Burger Kings (the more usual places of consumption that offered cheap minimal nutrition). The demands of the labor process (often late and eratic hours) in relation to restricted locational choices for living (given rents, housing affordability, public transport availability – car ownership is not feasible, and the like) reinforced geographical segregation. The insertion of racially marked and gendered bodies into this system trapped certain social groups into the dead-end prospects associated with these impoverished zones (see Fernandez-Kelly, 1994; more generally, Hanson and Pratt, 1994).

It is hard to do justice to the appalling effects of such conditions at all points in this particular process of circulation variable capital. Lack of respect and dignity in the workplace, negligible bargaining power in the labor market, minimal and health-threatening forms of consumption and terrible conditions of child-rearing are characteristic. The marks of all this violence upon individual bodies are not hard to read. Systematic studies again and again emphasize the stark impacts of inequalities upon life chances. Baltimore City has the lowest life expectancy of almost any other comparable jurisdiction in the United States (and comparable to many impoverished and undeveloped countries). 'In the groups we studied,' write Geronimus et al. (1996, 1555–6), after a comparative study of similar zones of Detroit, New York City, Los Angeles, and Alabama, 'the number of years of life lost generally increased with the percentage of people in the group who were living in poverty, with the poverty rate accounting for more than half the racial differences in mortality.' The data tell an appalling story: 'the probability that a 15 year old girl in Harlem would survive to the age of 45 was the same as the probability that a typical white girl anywhere in the United States would survive to the age of 65.' While it would be wrong to argue that lack of a living wage is the only factor at work here, the associations are far too strong to deny an active connection.

A campaign for a 'living wage,' organized by Baltimoreans United in Leadership Development (BUILD) seeks to change all this. BUILD was founded in 1978, through the coming together of the Interfaith Ministerial Alliance (predominantly though by no means exclusively African-American) that had been an important church-based force for civil rights with the Industrial Areas Foundation (IAF, a Chicago-based Saul Alinsky style community empowerment organization). BUILD became an activist voice for social change and economic development in the city dedicated to the improved well-being of impoverished and marginalized populations. It played an important role in struggles to regenerate failing neighborhoods and it initially joined wholeheartedly in the city and corporate-led strategy to generate employment through public investments and subsidies to business (as, e.g., in the Inner Harbor renewal, the construction of a convention center, a new ballpark, etc., all in the downtown core).

In the early 1990s, BUILD recognized that its strategies were too limited. Revitalized neighborhoods lacking adequate employment slipped back into decay. The public investment and subsidies to corporations were producing below-poverty jobs. The corporate-backed revitalization of downtown had not delivered on its promises and was increasingly viewed by BUILD as a 'great betrayal.' The churches that formed the basis of BUILD found themselves pushed to deliver more and more in the way of social services (soup kitchens, clothing, social assistance) to a population for whom Groucho Marx's witticism – 'Look at me, I've made my way up from nothing to a state of extreme poverty' – was cruel as well as a joke.

Consistent with its religious roots, BUILD decided to launch a campaign in the name of 'family values' and 'community' betterment, for a 'living wage.' They argued that business, in return for public subsidies, should commit itself to a social compact. This translated into the ideal of a minimum wage of $7.70 per hour, permanent jobs, adequate benefits, and career opportunities for all workers. Recognizing the difficulty of achieving this overnight, BUILD proposed an immediate wage hike to $6.10 an hour rising to $6.60 in July 1996 and going to $7.10 in 1997 and $7.70 in 1999. This is actually a minimalist demand (it is worth noting that the most recent piece of living wage legislation in San Jose, California, set the level at $10.75).

Like all such struggles, as Marx observed (1976 edition, 409), the role of 'allies in those social layers not directly interested in the question' is of considerable significance. The impetus for the campaign came from the churches. This set the tone concerning the definition of moral and civilized behavior that always enters into the determination of the value of labor power. What BUILD in effect says is that the market valuation of labor power as it now occurs in Baltimore is unacceptable as a 'moral' datum for a 'civilized' country. The focus on jobs connected immediately to the institutions of labor. A new form of labor organizing was needed which drew upon the skills of IAF, the power of AFSCME (State, County and Municipal Employees, which became a full partner in the campaign in 1994, providing personnel and resources). This meant a move away from traditional workplace industrial organizing towards a city-wide movement to change the baseline conditions for the circulation of variable capital. Jonathan Lange (1996), the labor organizer working with BUILD, outlines the strategy as follows:

> Organizing is a relational activity, it takes place *in* a place among people, and it is not totally mobile like capital. Ultimately you are not organizing workplaces and factories you are organizing people so ... the industrial model does not make total sense. So you've got to figure out how to organize ... a total labor market no matter where people work, to build an organization that is transportable for people from workplace to workplace, which means that the benefit plans have to be portable, the relationships in the organization have to be portable and not built all totally on one work place, which means that you have to understand people are not going to be leaders necessarily right away but potential leaders who can develop a following in their current workplace or when they move into their new one. It means you have to target those industries and corporations where your ability to withhold labor isn't the only strength you have, that you have other sorts of ways of getting leverage to try and reach recognition and accommodation ... This is an experiment to try to figure out whether within a

certain labor market if you merge, if you ally working people with other
kinds of decency and power and you carefully target institutions that are not
totally mobile, that cannot just run away with their capital, can workers get
themselves on a more equal footing? And if you do that enough ... can you
begin to really raise the basis, the floor of wages in a city?

The strategy is, then, two pronged. First, build a cadre of workers who can
carry their leadership skills and potentialities with them. Some workers –
mostly African-American women and men – immediately joined up to lead
a Solidarity Sponsoring Committee that adopted as its motto 'Climbing
Jacob's Ladder.' But others were more reluctant. Second, push hard to
create a powerful alliance of forces to change the baseline for the circulation
of variable capital. Initially, BUILD's strength lay in the churches. But the
fact that it was mainly women and African-American women who were
suffering conjoined questions of gender, race, and class in ways that could
potentially unify a variety of social movements (including the unions as
well as civil rights and women's organizations). The campaign, moreover,
made great play with the concept of the dignity of labor and of the laborer,
even daring to argue sometimes that the rule that 'any job was better than
none' ought to be brought into question when the quality, potentiality, and
dignity of available labor processes was taken into account.

The campaign won significant concessions in 1995. City Hall now
mandates that all city wages and all sub-contracts with the city should
honor the 'living-wage' policy. Though the Mayor initially resisted on the
grounds of keeping Baltimore competitive in the face of 'globalization,' he
now claims the effort is cost-effective (when the reduced cost of social
services to the impoverished poor is factored in). The World Trade
Center (run by the State Government) has followed suit (with, interestingly, support from the business tenants in the State-operated building
but heavy criticism from business leaders in the State). Early in 1998, the
City School Board agreed to a living-wage clause in all its subcontracts.
Now the Johns Hopkins System is faced with exactly that same question,
both as the supplier of services (through Broadway Services) and, being
the largest private employer in the State, as a demander of them (an
interesting example of how capital so frequently operates on both sides of
the supply-demand equation when it comes to labor – cf. Marx's argument, 1976 edition, 752). To this end a campaign began early in 1996 to
persuade the Johns Hopkins System to accept the living wage as part of its
own contractual practices.

The search for allies within the Johns Hopkins System became crucial.
The Graduate Representative Organization together with some faculty
and, ultimately, the Black Student Union and some representatives of the

student council took up the question. Initially there was also a surprising degree of indifference, even on the part of campus groups that ought to have been immediately interested in the question. Some economists in the University argued (rather predictably) against any interference in free-market forces, on the grounds 'that most people earning the present minimum wage are worth just that' (Hanke, 1996). Plainly, the outcome of the struggle depended (and continues to depend) not only on the capacities of the Solidarity Sponsoring Committee (SSC) (with AFSCME's help) to organize and the powers of moral suasion of BUILD but also upon the ability to create a powerful alliance within Johns Hopkins itself behind the idea that a living wage is mandatory for all those who work directly or indirectly (through sub-contracts) within the institution. By 1998, most students and most faculty were persuaded of the idea but were still faced by a recalcitrant administration. By 1999, the latter, in response to both internal and external pressures (both financial and moral), had tardily recognized its responsibilities towards the appalling conditions of impoverishment and ill-health that predominated in its shadow. It also finally acknowledged that its own wage policies might have some role in the construction of such conditions. It announced it would 'become a leader' among the universities on the living wage issue and ensure that everyone would receive at least $7.75 an hour (the 1996 living wage) by 2002.

The Baltimore campaign for a living wage (which is currently being replicated in some thirty or so other cities as well as at the state level elsewhere – see Pollin and Luce, 1998) offers a rather special set of openings to change the politics of how bodies are constructed/destroyed within the city. Its basis in the churches, the community, the unions, the universities, as well as among those social layers 'not immediately concerned with the question,' starts to frame body politics in a rather special way, by-passing some of the more conventional binaries of capital/labor, white/black, male/female, and nature/culture. Radical social constructionists should presumably relish rather than frown upon this confusion of terms. If, for example, Butler's (1993, 9) argument for 'a return to the notion of matter, not as site or surface, but as a process of materialization that stabilizes over time to produce the effect of boundary, fixity, and surface we call matter,' is taken as the proper framing for understanding the body in a situation of this sort, then the 'living wage' campaign is a fundamental form of body politics. This is not to say its mode is unproblematic. Consistent with its religious roots and its emphasis upon a traditional conception of the family as a proper unit of reproduction, the religious side of the campaign could be viewed as or even turn exclusionary. And BUILD in general seeks its own empowerment as a political organization as well as the empowerment of the low-income population it seeks to serve. Yet these are not reasons to abjure the living-wage objective. In practice many different interests (some secular as well as religious) now support the common goal of a decent living wage for everyone who works in Baltimore.

The 'living-wage' issue is fundamentally a class issue that has ramifications across the moments of production, exchange, and consumption. It has the power, therefore, to define what the 'work' side of current proposals for 'workfare' welfare reform might be about. Unfortunately, this potential relationship is now being inverted as the city is forced to absorb several thousand (possibly as many as 14,000) workfare recipients into its labor force (the total employment in all categories downtown is around 100,000). Both the city and Johns Hopkins began to employ workfare recipients at $1.50 an hour (as 'trainees'), and in the first rush this meant some displacement of minimum wage workers. The effect was to create an even lower datum than that set by the legal minimum wage for the circulation of variable capital within the city. A political struggle organized by BUILD citywide and a coalition of forces within Johns Hopkins led to the commitment by the Governor and by the President of the Johns Hopkins that there would be no displacement of existing workers by workfare trainees.

This is not an easy political battle to win more generally and its unfolding is illustrative of how class struggle gets waged from the capitalist side. Burger King, for example, has one of its most profitable franchises in Baltimore. Located in an 'empowerment zone' it is eligible for government subsidies and it can employ workers off the welfare rolls as 'trainees' at a cost far below the minimum, let alone the living, wage. Yet Burger King gets cited by President Clinton in his 1997 State of the Union Address as one of the large companies willing to hire people off the welfare rolls, and the President promised to press for special tax credits for companies that did this. Later, however, under strong pressure from organized labor and many community groups around the country, the President agreed (against intense Republican opposition) to bring all workfare employment within the framework of labor laws (allowing organizing of workfare workers and protection from the grosser forms of direct exploitation). Thus does the accumulation of capital proceed, with state assistance mainly going to capital, as class struggle unfolds around one of the most contested and fraught social issues of the 1990s in the United States.

The living-wage campaign integrates race, gender, and class concerns at the level of the 'city' as a whole. In particular, it opens up potential leadership roles for African-American women to alter bodily practices and claim basic economic rights. The campaign furthermore proposes a different spatial model of political intervention in the valuation of labor power, highlighting Munn's argument that 'bodily spacetime serves as a condensed sign of the wider spacetime of which it is a part' (1985, 17). Creating an alternative spatial frame to that of increasingly fragmented workplaces (within which the value of labor power can only be established piecemeal) becomes part of the means to alter the conditions of circulation of variable capital. The campaign offers the possibility for broad-based coalition politics at a different spatial scale.

Changing the baseline conditions of the circulation of variable capital will not change everything that needs to be changed in Baltimore either within the labor process or without. It will not automatically improve the quality of the work experience. It does not automatically confront the sexual harassment of the women on the job, the rampant racism in the city, manifestations of homophobia, the dramatic deterioration of many Balti-more neighborhoods, or even the stresses within and around the institu-tion of the family. Nor does it open the door to revolution rather than reform of the wage system (abolition of the wages system is hardly an issue here whereas the reformist claim – of which Marx was roundly critical – for a fair day's wage for a fair day's work is). But it does create necessary conditions for the transformation of bodily practices on the part of a sub-stantial number of working people in Baltimore. Without that, many other possibilities for social transformation are blocked. Marx (1967 edition, Volume 3, 320), recognizing the dilemma, put it this way in a remarkable passage that deserves some thought:

[T]he realm of freedom actually begins only when labour which is determined by necessity and mundane considerations ceases; thus in the very nature of things it lies beyond the sphere of actual material production. Just as the savage must wrestle with Nature to satisfy his wants, to maintain and reproduce his life, so must civilized man, and he must do so in all social formations and under all possible modes of production. With his develop-ment this realm of physical necessity expands as a result of his wants; but, at the same time, the forces of production which satisfy those wants also increase. Freedom in this field can only consist in socialized man, the associated producers, rationally regulating their interchange with Nature, by bringing it under common control, instead of being ruled by it as by the blind forces of Nature; and achieving this with the least expenditure of energy and under conditions most favourable to, and worthy of, their human nature. But it nonetheless still remains a realm of necessity. Beyond it begins that development of human energy which is an end in itself, the true realm of freedom, which, however, can blossom forth only with the realm of necessity as its basis. The shortening of the working day is its basic prerequisite.

To that remarkable passage with its startlingly reformist last sentence we can also add: 'an adequate living wage is likewise a basic prerequisite.' The struggle for a living wage within the space of Baltimore has its place in a more universal struggle for rights, for justice, dignity, and decency in all the interstices of a globalizing capitalism. Its particularities make it peculiar, give it strengths and weaknesses, but they are not irrelevant to the achievement of a more universalizing politics. And while the numbers of people so far affected are small, the manner of these campaigns illustrates how frustration of politics at one scale can potentially be met by a shift to a different scale of political action.

3 Bodies in space and time

The body that is to be the 'measure of all things' is itself a site of contestation for the forces that create it. The body (like the person and the self) is an internal relation and therefore open and porous to the world. Unfortunately the relational conception of the body can all too easily take an idealist turn, particularly in academic politics. The body is not monadic, nor does it float freely in some ether of culture, discourses, and representations, however important these may be in materializations of the body. The study of the body has to be grounded in an understanding of real spatio-temporal relations between material practices, representations, imaginaries, institutions, social relations, and the prevailing structures of political-economic power. The body can then be viewed as a nexus through which the possibilities for emancipatory politics can be approached. While there are some remarkable insightful writings on that theme available to us, it is worthwhile remembering the vital insights to be had from Marx's understanding of how bodily materializations occur within the circulation of capital under capitalist social relations. The body may be 'an accumulation strategy in the deepest sense' but it is also the locus of political resistance given direction, as the example of BUILD's campaign for a living wage in Baltimore illustrates, by the basic fact that we are, in the most literal sense, political animals rendered capable of moral argument and thereby endowed with the capacity to transform the social relations and institutions that lie at the heart of any civil society. Laborers are, in short, positioned to claim rights consistent with notions of dignity, need, and contribution to the common good. If those claims are unrealizable within the circulation of variable capital then, it seems, the revolutionary demand to escape such constraints is a fundamental aspect of what body politics must be about. We shall need to consider it.

PART 3

THE UTOPIAN MOMENT

CHAPTER 8

The spaces of Utopia

A map of the world that does not include Utopia is not even worth glancing at.

(Oscar Wilde)

1 The Baltimore story

I have lived in Baltimore City for most of my adult life. I think of it as my home town and have accumulated an immense fund of affection for the place and its people. But Baltimore is, for the most part, a mess. Not the kind of enchanting mess that makes cities such interesting places to explore, but an awful mess. And it seems much worse now than when I first knew it in 1969. Or perhaps it is in the same old mess (see Table 8.1) except that many then believed they could do something about it. Now the problems seem intractable.

Too many details of the mess would overwhelm. But some of its features are worth pointing out. There are some 40,000 vacant and for the most part abandoned houses (Plate 8.1) in a housing stock of some 300,000 units within the city limits (there were 7,000 in 1970). The concentrations of homelessness (in spite of all those vacant houses), of unemployment, and, even more significant, of the employed poor (trying to live on less than $200 a week without benefits) are everywhere in evidence. The lines at the soup kitchens (there were 60 of them in the State of Maryland in 1980 and there are now 900) get longer and longer (30 percent of those using them have jobs according to some informal surveys) and the charity missions of many inner-city churches are stretched beyond coping (Plate 8.2). The inequalities – of opportunities as well as of standards of life – are growing by leaps and bounds. The massive educational resources of the city (Baltimore City has some of the finest schools in the country, but they are all private) are denied to most of the children who live there. The public schools are in a lamentable state (two and a half years behind the national average in reading skills according to recent tests).

Table 8.1 Then and now: An inner-city Baltimore neighborhood

	1966	1988
Economic percentages		
Adult unemployment rate	7.0	17.0
Households receiving welfare	28.0	30.0
Households with incomes less than $10,000 (1988 dollars)	41.0	47.0
Households with incomes less than $20,000 (1988 dollars)	16.0	18.0
Adults who are high school graduates	10.0	49.0
Households in which at least one person owns a car	23.0	36.0
Percentage employed as laborers	43.0	8.0
Percentage doing clerical work	1.0	30.0
Household and family structure		
Median household size	2.9	1.9
Percentage of adults retired	13.0	30.0
Percentage of population under 18 years of age	45.0	34.0
Percentage of households with children with a male adult	56.0	43.0
Percentage of one-person households	16.0	31.0
Percentage of households with five or more people	30.0	12.0
The neighborhood		
Most commonly cited 'good' aspect	people	people
Most common complaint	housing	drugs/ crime
Percentage of residents who are renters	85.0	78.0
Percentage of adults living in neighborhood more than 10 years	48.0	60.0
Percentage who think neighborhood is improving	n.a.	14.0

The data above were compiled by surveys commissioned by the Baltimore Urban Renewal and Housing Agency in 1966 and repeated by the *Baltimore Sun* in April 1988. They focus on a neighborhood most seriously affected by the uprisings that followed the assassination of Martin Luther King. They were published in the *Baltimore Sun* (April 4th, 1988) on the twentieth anniversary of that event.

With the exception of high school graduation rates, car ownership, and over-crowding, the data indicate at best stability and in some instances a worsening of economic and social conditions in the neighborhood. While some of this is due to an aging population, the lack of employment opportunities is evident. The catastrophic drop in laboring employment and the rise of clerical work conceals a radical shift in employment opportunities from men to women. This correlates with much higher levels of high school education as well as the drop in male adult presence in households. Data compiled during the 1990s for similar neighborhoods give no evidence of any reversal of these trends.

(Source: Harvey, 1988, 238)

*Plate 8.1 **Abandonment of the city: housing in Baltimore**. In 1970 there were circa 7,000 abandoned houses in Baltimore City. By 1998 that number had grown to an estimated 40,000 out of a total housing stock of just over 300,000 units. The effect on whole neighborhoods has been catastrophic. City policy is now oriented to large scale demolition (4,000 were demolished between 1996 and 1999 and another 11,000 demolitions are planned). The 'official' hope is that this will drive the poor and the underclass from the city. The idea of reclaiming older neighborhoods – particularly those with a high quality housing stock – for impoverished populations has been abandoned even though it could make much economic and environmental sense.*

Plate 8.2 Charity in the city: Our Daily Bread in downtown Baltimore. Our Daily Bread run by Catholic Charities feeds around 900 people daily. Visited by the Pope, it has long been a flagship operation for servicing the inner city poor. But in 1998, the Downtown Partnership, led by Peter Angelos, the multi-millionaire owner of the Baltimore Orioles (salary budget for baseball players at $90 million annually), began to agitate against poor people circulating in the downtown area because they supposedly fostered crime, devalued properties and deterred redevelopment. The Partnership urged the city to set up a 'social services campus' for the poor away from the downtown area. Catholic Charities was asked to seek a less central location. In April 1999 it was announced that Our Daily Bread would be moved to a renovated building donated by Angelos, symbolically tucked away behind the city jail in an impoverished neighborhood. When local residents complained Catholic Charities abandoned that site and began looking elsewhere. The bourgeoisie, as Engels argued, have only one solution to social problems – they move them around while blaming those least able to deal with the burden.

Chronic poverty and all manner of signs of social distress (Plate 8.3) reign in the shadow of some of the finest medical and public health institutions in the world that are inaccessible to local populations (unless they have the privilege to clean the AIDS wards for less than a living wage or have medicare/medicaid status or a rare disease of great interest to elite medical researchers). Life expectancy in the immediate environs of these internationally renowned hospital facilities is among the lowest in the nation and comparable to many of the poorer countries in the world

Plate 8.3 Poverty in the city: in the shadow of Johns Hopkins hospital. The Johns Hopkins hospital and its associated School of Public Health are rated as among the best in the world. Yet the life-expectancy of individuals in the city is abysmally low and the health statistics in the immediate environs of these institutions tell an appalling story of impoverishment, marginalization, exploitation, and neglect. The pawnshops, the crumbling storefront churches, the bailbondsmen, all in the vicinity of the hospital, signify the social distress. But a crumbling mural expressing the desire to 'Climb Jacob's Ladder' out of misery to a condition of self-acceptance and reliance provides a glimmer of utopian desire. The living wage campaign in the city and in Johns Hopkins (with its slogan of 'Climbing Jacob's Ladder') give hope for one step up that ladder.

(63 years for men and 73.2 for women). The rate of syphilis transmission is the highest of any city in the developed world (according to WHO statistics) and there has been an explosion of respiratory diseases (more than doubling for all categories in the city between 1986 and 1996, according to data collected for the Environmental Protection Agency, but led by an astonishing increase in the asthma rate from around 8 to nearly 170 per 10,000 inhabitants). The only notable public health success recorded in the city is the dramatic curbing of TB infections. This happened by way of a public health commissioner who, having had military medical experience in Vietnam, saw fit to adapt the Chinese communist idea of 'barefoot doctors' to urban Baltimore and thereby bring the city's TB rate down within a decade from its unenviable position of worst in the nation to below the national average.

The affluent (black and white) continue to leave the city in droves (at a net rate of over a thousand a month over the last five years according to the Census Bureau) seeking solace, security, and jobs in the suburbs (population in the city was close to a million when I arrived and is now down to just over 600,000). The suburbs, the edge cities, and the ex-urbs proliferate (with the aid of massive public subsidies to transport and upper-income housing construction via the mortgage interest tax deduction) in an extraordinarily unecological sprawl (Plate 8.4) – long commutes, serious ozone concentrations in summer (almost certainly connected to spiraling respiratory ailments), and loss of agricultural land. Developers offer up this great blight of secure suburban conformity (alleviated, of course, by architectural quotations from Italianate villas and Doric columns) as a panacea for the breakdown and disintegration of urbanity first in the inner city and then, as the deadly blight spreads, the inner suburbs. And it is there, in that bland and undistinguished world, that most of the metropolitan population, like most other Americans who have never had it so good, happily dwell. Residency in this commercialized 'bourgeois utopia' (as Robert Fishman, 1989, calls it) anchors the peculiar mix of political conservatism and social libertarianism that is the hallmark of contemporary America.

There has been an attempt of sorts to turn things around in the city. Launched in the early 1970s under the aegis of a dedicated and authoritarian mayor (William Donald Schaeffer) it entailed formation of a private-public partnership to invest in downtown and Inner Harbor renewal (Plate 8.5) in order to attract financial services, tourism, and so-called hospitality functions to center city. It took a lot of public money to get the process rolling. Once the partnership had the hotels (Hyatt got a $35 million hotel by putting up only half a million of its own money in the early 1980s), it needed to build a convention center to fill the hotels and

Plate 8.4 Bourgeois utopia: suburban sprawl. *Like many other metropolitan regions in the United States, Baltimore has exploded outwards at an extraordinary rate (see Plate 3.1). Impelled by a complex mix of fears of the city, compounded by racism and class prejudice, the collapse of public infrastructures in many parts of the city, and attracted by the 'bourgeois utopian' desire to secure isolated and protected comforts, the effect of this propertied individualism has been to create a remarkably repetitive landscape of low-density sprawl coupled with total dependence on the automobile. The ecological impacts are strongly negative and the social and economic costs of traffic congestion and infrastructure provision are rising rapidly.*

Plate 8.5 Developers' utopia: Baltimore's Inner Harbor renewal. Almost everything to be seen on the present skyline of Baltimore's Inner Harbor has been constructed since around 1970. The background buildings largely represent office and hotel spaces with high rise condominiums (both of which proved hard to sell off except at cut-prices) guarding either end. The tall condominium on the left was built on valuable land 'given away' to the developer in return for promises of help elsewhere that never materialized. In the foreground are the leisure and tourist activities that focus on the harbor front (Rouse's investments in a series of Pavilions occupy the central corner of the harbor). Built through a 'public-private partnership' much of the development has had a checkered history. The Hyatt Regency Hotel (center top) gave Hyatt a $35 million hotel for an investment of $500,000 (the rest was public moneys). While this investment eventually turned out successfully for the city, the Columbus Science Center (with the white fluted roofline center bottom) cost $147 million of publicly secured private moneys but its main function, a Hall of Exploration, was forced to close in 1997 after nine months of operation. Rescued from bankruptcy by a State takeover, the building is now run by the University of Maryland with a marine biotechnology center as a main tenant.

get a piece of what is now calculated to be an $83 billion a year meetings industry. In order to keep competitive, a further public investment of $150 million was needed to create an even larger convention center to get the big conventions. It is now feared that all this investment will not be profitable without a large 'headquarters hotel' that will also require 'extensive' public subsidies (maybe $50 million). And to improve the city image, nearly a half billion dollars went into building sports stadiums (Plate 8.6) for teams (one of which was lured from Cleveland) that pay several million a year to star players watched by fans paying exorbitant ticket prices. This is a common enough story across the United States (the National Football League – deserving welfare clients – calculates that $3.8 billion of largely public money will be poured into new NFL stadiums between 1992 and 2002). The state spends $5 million building a special light rail stop for the football stadium that will be used no more than twenty days a year.

This is what is called 'feeding the downtown monster.' Every new wave of public investment is needed to make the last wave pay off. The private-public partnership means that the public takes the risks and the private takes the profits. The citizenry wait for benefits that never materialize. Several of the public projects go belly up and an upscale condominium complex on the waterfront (Plate 8.7) does so poorly that it gets $2 million in tax breaks in order to forestall bankruptcy while the impoverished working class – close to bankruptcy if not technically in it – get nothing. 'We have to be competitive,' says the Mayor and that 'if they fail then no one else will want to invest,' apparently forgetting that the higher tax bills on the rest of us (including those who might upgrade their properties) is also an incentive to join the exodus from the city to the suburbs that has long been under way.

There is, of course, a good side to the renewal effort. Many people come to the Inner Harbor. There is even racial mixing. People evidently enjoy just watching people. And there is a growing recognition that the city, to be vibrant, has to be a twenty-four-hour affair and that mega bookstores and a Hard Rock cafe have as much to offer as Benetton and the Banana Republic (Plate 8.8). A hefty dose of social control is required to make such activities viable and signs of such control are omnipresent (Plate 8.9). The wish to be close to the action brings some young professionals (those without kids) back into center city. And when 'gentrification' in the classical sense of displacement of low-income populations has occurred (as it has mainly around the harbor) it has at least physically revitalized parts of the city that were slowly dying from neglect (Plate 8.10). Some of the seedier public housing blocks have been imploded to make way for better quality housing in better quality environments (Plate 8.11). Here and there, neighborhoods

Plate 8.6 Public investments in the city: stadiums and a convention center for the affluent. During the 1990s nearly a billion dollars went into two publicly-financed sports stadiums ($500 million), an extension to the Convention Center ($150 million) and other major downtown projects (e.g. the addition of a light rail stop for the football stadium to be used no more than twenty times a year for $5 million). The argument for such investments is that they create jobs and generate income. But a careful cost-benefit analysis by two respected economists (Hamilton and Kahn, 1997) showed a net loss of the baseball stadium investment of $24 million a year. Meanwhile, libraries have been closed, urban services curtailed and investment in city schools has been minimal.

Plate 8.7 Public subsidy and private gain: the story of Harborview. After the Key Highway Shipyard closed in 1982 (with the loss of 2,000 jobs), the vacant site (top) became a focus of lengthy controversy. Approval was finally given in 1987 to build a series of high rises on the site, in the face of fierce local community opposition because the sheer scale of the project threatened the intimacy of existing neighborhoods and because access to the waterfront would be compromised. Funding for the project, initially confused by a mortgage foreclosure and multiple transfers of developer rights, was finally (and abruptly) procured from southeast Asia (Parkway Associates, then awash with surplus funds, put up the money without question since the site reminded their agent of Hong Kong). The project immediately hit difficulties with the financial crash of October 1987 and never seems to have turned a profit after the first tower was opened to much fanfare ('a new style in urban living') in 1993 (penthouse apartments marketed for $1.5 million). Eventually bailed out by a $2 million tax relief package in 1998, the developers have thrashed around to find ways to make the site more profitable. Proposals included building three more towers to make the first tower more viable. In 1999, construction began on luxury town houses and 'canal homes' with some modest high rise construction layered in between on the landward side. Another tower may yet be built.

Plate 8.8 Degenerate utopia in the city – the urban spectacle as a commodity. In the wake of the urban unrest that shook the city in the 1960s, an influential elite of government officials and business leaders sought to rescue downtown investments by pursuit of consumerism and tourism. The urban spectacle constructed around the Inner Harbor is now reputed to draw more visitors to Baltimore than Disneyland. Rouse's pavilions at Harbor Place (top right) provide the anchor, but the general scene of leisured consumerism has its institutional elements (the National Aquarium and the Maryland Science Center), its interior version (Rouse's Gallery at Harbor Place, left) and its more recently added eternal symbols such as a Hardrock Cafe, an ESPN Zone, and Planet Hollywood (bottom right).

Plate 8.9 Degenerate utopia in the city – spectacle and social control at the Maryland Science Center. The Maryland Science Center presents two faces to the world. Opening onto the commercialism of the inner harbor, the Center beckons as a friendly space in which we can learn and, on this occasion, experience 'Videotopia' (at a price). But the backside of the building tells another story. It was opened in 1976 as one of the first pieces in the inner harbor revitalization. It looks like a fortress from the back. Initially it had no entrance facing the community or even onto the street. The building was designed in the wake of the 1968 unrest that followed the assassination of Martin Luther King. The African-American community that then lived nearby (largely displaced since by highway construction and gentrification) was perceived as a threat. So the fortress design was deliberate. It was meant to repel social unrest and function as a strategic (bunker-style) outpost at the south end of the inner harbor to protect the investments yet to come.

Plate 8.10 Yuppie utopia: gentrification and renewal in the Canton district of Baltimore. The successful recycling of older industrial buildings (America Can, left) and new townhouses on the nearby waterfront (top right) have led the way in the rapid gentrification of the Canton neighborhood on the eastern side of the city. Within sight of downtown, the strip running from Canton to the city center along the harbor edge is known as 'The Gold Coast' because of its potential for up-scale redevelopment. The effect on older housing in Canton has been startling. Lacking any other space to expand, the owners of narrow traditional row-houses compete by building bizarrely conspicuous roof decks with views over the harbor (bottom right).

Plate 8.11 Rehousing the poor. *The public housing (top) largely constructed in the 1950s and 1960s needed renewal and was in any case increasingly blamed as a negative environment that fostered criminality and other forms of anti-social behavior. Imploded in the 1990s, it has been replaced with low-rise suburban-style architecture in a gated community atmosphere within sight of downtown. 'Pleasant View Gardens' (bottom) is now claimed as an example of the 'new urbanism' – a village style gated space in an inner city setting.*

have pulled themselves together and developed a special sense of community that makes for safer, more secure living without degenerating into rabid exclusionism. In a few neighborhoods major projects have been launched, using an array of public and private resources, to revitalize impoverished communities (Plate 8.12). None of this touches the roots of Baltimore's problems.

One of those roots lies in the rapid transition in employment opportunities. Manufacturing jobs accelerated their movement out (mainly southwards and overseas) during the first severe post-war recession in 1973–5 and have not stopped moving since (see Table 8.2). Shipbuilding, for example, has all-but disappeared and the industries that stayed have 'downsized.' Bethlehem Steel (Plate 8.13) employed 30,000 in 1970; less than 5,000 now make nearly the same amount of steel with round after round of high tech investment, the last of which received a $5 million state subsidy. General Motors – another deserving welfare client – received a massive Urban Development Action Grant in the early 1980s to keep its assembly plant open, and now threatens to close its truck assembly operation. City and state representatives are scrambling to find a sufficiently lucrative aid package to keep them in town. Containerization of port operations and automated ship loading (Plate 8.13) have reduced employment on the docks to a shadow of its former importance.

Service jobs have materialized to replace perhaps as many as a quarter of a million jobs lost in manufacturing and port operations. Within the city many of these are low paying (with few benefits), temporary, non-unionized and female (Plate 8.14). The best many households can hope for is to keep their income stable by having two people work longer hours at a lower individual wage. The general absence of adequate and affordable day care means that this does not bode well for the kids. Poverty entraps and gets perpetuated, notwithstanding the campaign for a 'living wage' that struggles to improve the lot of the working poor and protect the many thousands now being pushed off welfare into a stagnant labor market (see Chapter 7). Conversion of older industrial facilities here and there provides new sources of livelihood that offer some support for neighborhood revitalization.

The geographical disparities in wealth and power increase to fashion a metropolitan world of chronically uneven geographical development. For a while the inner suburbs drained wealth from the central city but now they, too, have 'problems' though it is there, if anywhere, where most new jobs are created. So the wealth moves, either further out to ex-urbs that explicitly exclude the poor, the underprivileged, and the marginalized, or it encloses itself behind high walls, in suburban 'privatopias' and urban 'gated communities' (Plate 8.15). The rich form ghettoes of affluence (their

Plate 8.12 Neighborhood revitalization: Sandtown-Winchester and the James Rouse paradox. Wholesale community renewal, with a mix of support from public and private sources, has been attempted in Sandtown-Winchester. Actively pushed by a community organization (BUILD) it was soon recognized that lack of jobs and low wages lay at the heart of community degradation. The 'living wage campaign' (see Chapter 7) followed on from the Sandtown-Winchester experience. James Rouse participated in the Sandtown-Winchester effort through his Enterprise Foundation (set up after his retirement to assist in the revitalization of poor neighborhoods). Having, through his activities as a largely suburban and commercial developer, helped destroy the viability of inner city living, he returned after his retirement to try to revitalize areas that his own activities had done so much to destroy. Rouse's work is now commemorated by a controversial mural in Sandtown-Winchester that makes him appear as a benevolent patriarch on an African-American plantation.

Table 8.2   Where the jobs have gone: Baltimore 1980–5

Company	Type of business	Number of jobs lost
*Acme Markets	Grocery Chain	1,200
*Airco Welding	Cored Wire	150
*Allied Chemical	Chromium	145
Bethlehem Steel	Steel	7,000
*Bethlehem Steel Shipyard	Ship Repair	1,500
*Brager-Gutman	Retail	180
*Cooks United	Discount Stores	220
Esskay	Meat Packing	240
General Electric	Electrical Goods	550
General Motors	Auto Parts	247
*Korvettes	Department Stores	350
*Maryland Glass	Glass	325
*Maryland Shipping and Dry Dock	Ship Repair	1,500
Max Rubins	Apparel	225
*Misty Harbor Raincoat	Rainwear	210
*Pantry Pride	Grocery Chain	4,000
*Plus Discount Stores	Discount Stores	150
*Two Guys	Discount Stores	150
Vectra	Fiber and Yarn	600
*Western Electric	Electrical Production	3,500

The recession of 1980–5 brought another powerful wave of job losses to the Baltimore region, as illustrated in this list of cutbacks. The list is adapted from the chart appearing in the *Baltimore Sun*, March 21st, 1985. Companies marked with an asterisk shut down operations completely. In the case of such shutdowns, there had often been a significant run down of the labor force in earlier years. The jobs lost in retailing were eventually regained but those in manufacturing were permanently lost. This period saw the end, for example, of shipbuilding and ship repair as a staple of the Baltimore economy.

(Source: Harvey, 1988, 236)

'bourgeois utopias') and undermine concepts of citizenship, social belonging, and mutual support. Six million of them in the United States now live in gated communities as opposed to one million ten years ago (Blakely, 1997). And if communities are not gated they are increasingly constructed on exclusionary lines so that levels of segregation (primarily by class but also with a powerful racial thread) are worse now in Baltimore than ever.

The second major root of the mess lies in institutional fragmentation and breakdown. City Hall, caught in a perpetual fiscal bind buttressed by the belief that slimmer government is always the path to a more competitive city, reduces its services (while increasing its subsidies to corporations) whether needed or not. The potential for cooperation with

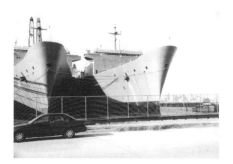

*Plate 8.13 The deindustrialization of Baltimore. Baltimore lost two-thirds of it manu-
facturing employment after 1960 (a net loss of 100,000 jobs or so). In World War II, for
example, it built nearly 500 'Liberty Ships' but the only residual sign of that activity today is
caring for some mothballed ships (middle left). Derelict plants litter the landscape waiting for re-
use (the abandoned brewery, middle right). The industry that remains, such as Bethlehem Steel
(bottom left and right) offer much reduced employment opportunities. The plant that employed
nearly 30,000 workers in 1970 now employs less than 5,000, leaving the empty parking lots to the
seagulls. Containerization (top right) and automation of the port (coal-loading, top left) have
reduced employment there also.*

Plate 8.14 The temporary worker.
The loss of manufacturing employment
and blue-collar unionized jobs has been
offset by increasing employment in ser-
vices (such as health care, making the
Johns Hopkins system the largest private
employer in the State of Maryland),
finance, insurance and real estate, aug-
mented by strong growth in the 'hospi-
tality industry' associated with the
convention trade and tourism. But many
of the new jobs are temporary, low paid,
with few if any benefits attached. And
women are much in demand.

suburban jurisdictions is overwhelmed by competitive pressures to keep taxes down, the impoverished and marginalized out, and the affluent and stable in. The Federal Government decentralizes and the State, now dominated by suburban and rural interests, turns its back on the city. Special tax-assessment districts spring up so that neighborhoods can provide extra services according to their means. Since the means vary, the effect is to divide up the urban realm into a patchwork quilt of islands of relative affluence struggling to secure themselves in a sea of spreading squalor and decay. The overall effect is division and fragmentation of the metropolitan space, a loss of sociality across diversity, and a localized defensive posture towards the rest of the city that becomes politically fractious if not downright dysfunctional.

The Downtown Partnership, to take one example, is led by Peter Angelos, the richest lawyer in the state and owner of the Baltimore Orioles. (He began his career as lawyer for the steelworkers dealing with occupational health and safety issues and made millions on the asbestos claims that bankrupted several major corporations as well as many of the 'names' of Lloyds in London that were foolish enough to insure them.) The Partnership is taking over the downtown, seeking to push the homeless – and the soup kitchens (particularly Our Daily Bread, Plate

Plate 8.15 Privatopias: the gated communities of Baltimore. More and more people in the United States choose to live in the protected spaces of gated communities. Baltimore is no exception and more and more such communities are being built within the city limits as well as in the suburbs.

8.2) that attract them – towards some peripheral zone. It even proposes a ghettoized 'campus for the homeless,' somewhere out of sight. City Hall falls into line and seeks extensive demolition of whole blocks of low-income housing, hoping to force the poor out into the suburbs, in perfect alignment with Engels's observation from long ago that the only solution the bourgeoisie can ever find for its problems is to move them around.

The prospects for institutional reform seem negligible. A tangled mix of bureaucratic and legal inflexibilities and rigid political institutional arrangements create ossified urban governance. Exclusionary communitarianism, narrow vested interests (usually framed by identity politics of various sorts – predominantly racial at the populist level, though in Baltimore there is a good deal of ethnic rivalry thrown in), corporate profit hunger, financial myopia, and developer greed all contribute to the difficulties. New resources are built into the social, political, and physical landscape of the metropolitan region so as to exacerbate both the inequalities and the fragmentations (most particularly those of race). There is, it seems, no alternative except for the rich to be progressively enriched and the poor (largely black) to be regressively impoverished. If the latter misbehave, they can always be incarcerated in that other place of massive public investment, the new city correctional institution (Plate 8.16).

In the midst of all this spiraling inequality, thriving corporate and big money interests (including the media) promote their own brand of identity politics, with multiple manifestos of political correctness. Their central message, repeated over and over, is that any challenge to the glories of the free market (preferably cornered, monopolized, and state subsidized in practice) is to be mercilessly put down or mocked out of existence. The power of these ideas lies, I suspect, at the core of our current sense of helplessness. 'There is no alternative,' said Margaret Thatcher in her heyday. Even Gorbachev agreed. The corporatized media relentlessly and endlessly repeat the refrain. An overwhelming ideological configuration of forces has been created that will brook no opposition. Those who have the money power are free to choose among name-brand commodities (including prestigious locations, properly secured, gated, and serviced), but the citizenry as a whole is denied any collective choice of political system, of ways of social relating, or of modes of production, consumption, and exchange. If the mess seems impossible to change then it is simply because there is indeed 'no alternative.' It is the supreme rationality of the market versus the silly irrationality of anything else. And all those institutions that might have helped define some alternative have either been suppressed or – with some notable exceptions, such as the church – been brow-beaten into submission. We the people have no right to choose what kind of city we shall inhabit.

Plate 8.16 Public investments in the city: correctional facilities (gated communities) for the poor. The only investment of direct interest to the city's poor is the extension to the city correctional institutions. Public investment in prisons has been a major growth sector in the US economy during the 1990s (with more than a million people now incarcerated). At the same time, all other forms of welfare provision have been heavily cut (forcing 14,000 people off the welfare rolls in Baltimore City alone). It costs more than $25,000 a year to house one prisoner.

But how is that we are so persuaded that 'there is no alternative'? Why is it, in Roberto Unger's (1987a, 37) words, that 'we often seem to be (such) helpless puppets of the institutional and imaginative worlds we inhabit.' Is it simply that we lack the will, the courage, and the perspicacity to open up alternatives and actively pursue them? Or is there something else at work? Surely it cannot be lack of imagination. The academy, for example, is full of explorations of the imaginary. In physics the exploration of possible worlds is the norm rather than the exception. In the humanities a fascination with what is called 'the imaginary' is everywhere apparent. And the media world that is now available to us has never before been so replete with fantasies and possibilities for collective communication about alternative worlds. Yet none of this seems to impinge upon the terrible trajectory that daily life assumes in the material world around us. We seem, as Unger (1987a, 331) puts it, to be 'torn between dreams that seem unrealizable and prospects that hardly seem to matter.' Is it really a choice between 'Dreamworks' or nothing?

To be sure, the ideology and practices of competitive neoliberalism do their quietly effective and insidious work within the major institutions – the media and the universities – that shape the imaginative context in which we live. They do so with hardly anyone noticing. The political correctness imposed by raw money power (and the logic of market competition) has done far more to censor opinion within these institutions than the overt repressions of McCarthyism ever did. 'Possibility has had a bad press,' Ernst Bloch (1988, 7) remarks, adding 'there is a very clear interest that has prevented the world from being changed into the possible.' Bloch, interestingly, associated this condition with the demise, denigration, and disparagement of all forms of utopian thought. That, he argued, meant a loss of hope and without hope alternative politics becomes impossible. Could it be, then, that a revitalization of the utopian tradition will give us ways to think the possibility of real alternatives? Bloch (1986) clearly thought so.

Close to the center of Baltimore, in the Walters Art Gallery, hangs a painting entitled 'View of an Ideal City' (Plate 8.17). It portrays the idea of some long-ago dreamed of perfection of city form accredited, perhaps appropriately under the circumstances, to an unknown Italian artist of the late fifteenth century. I like to think it was painted as Columbus prepared to undertake his fateful voyage. Although its form and style are drawn from long ago, when hopes, fears, and possibilities were different, its spirit still burns bright within the heart of Baltimore as a reproachful commentary not only on the urban desolation outside the Gallery's walls but also upon the lack of visionary ideals with which to combat that desolation.

2 The figure of the city

The figures of 'the city' and of 'Utopia' have long been intertwined (see, e.g., Fishman, 1982, and Hall, 1988). In their early incarnations, utopias were usually given a distinctively urban form and most of what passes for urban and city planning in the broadest sense has been infected (some would prefer 'inspired') by utopian modes of thought. The connection long predates Sir Thomas More's first adventure with the utopian genre in 1516. Plato connected ideal forms of government with his closed republic in such a way as to fold the concepts of city and citizen into each other and the city state of Phaeacia depicted in Homer's *Odyssey* has many of the characteristics later alluded to by More. The Judeo-Christian tradition defined paradise as a distinctive place where all good souls would go after their trials and tribulations in the temporal world. All manner of metaphors flowed from this' of the heavenly city, the city of God, the eternal city, the shining city on a hill (a metaphor dearly beloved by President Reagan). If heaven is a 'happy place' then that 'other' place, hell, the place of 'the evil other,' cannot be far away. The figure of the city as a fulcrum of social disorder, moral breakdown, and unmitigated evil – from Babylon

Plate 8.17 View of the ideal city: from the Walters Art Gallery in Baltimore. This anonymous painting of the Central Italian School, done sometime towards the end of the 15th century, depicts a vision of an ideal city in the context of that time. It now decorates the walls of the Walters Art Gallery in Mount Vernon Place close to the center of Baltimore, the city where ideals have frayed and there appears to be no alternative.

and Sodom and Gomorrah to Gotham – also has its place in the freight of metaphorical meanings that the word 'city' carries across our cultural universe. Dystopias take on urban forms such as those found in Huxley's *Brave New World* or Orwell's *1984*. The word 'police' derives from the Greek 'polis' which means 'city.' If Karl Popper was right to depict Plato as one of the first great enemies of 'the open society' then the utopias that Plato inspired may just as easily be cast as oppressive and totalitarian hells as emancipatory and happy heavens.

It is hard to untangle the grubby day-to-day practices and discourses that affect urban living from the grandiose metaphorical meanings that so freely intermingle with emotions and beliefs about the good life and urban form. It is interesting to note how it is often at the geographical scale of small-scale city life that the ideals of utopian social orderings are so frequently cast. Plato put the maximum population at five thousand and 'democratic' Athens at its height probably had no more than six thousand participating 'citizens' (which did not, of course, include women or the many slaves). I cannot hope to untangle all such meta-phorical and symbolic meanings here. But we need to recognize their emotive power. A few illustrative connections may help consolidate the point that urban politics is fraught with deeply held though often subterranean emotions and political passions in which utopian dreams have a particular place.

'City air makes one free' it was once said. That idea took shape as serfs escaped their bonds to claim political and personal freedoms within the self-governing legal entities of medieval cities. The association between city life and personal freedoms, including the freedom to explore, invent, create, and define new ways of life, has a long and intricate history. Generations of migrants have sought the city as a haven from rural repressions. The 'city' and 'citizenship' tie neatly together within this formulation. But the city is equally the site of anxiety and anomie. It is the place of the anonymous alien, the underclass (or, as our predecessors preferred it, 'the dangerous classes'), the site of an incomprehensible 'otherness' (immigrants, gays, the mentally disturbed, the culturally different, the racially marked), the terrain of pollution (moral as well as physical) and of terrible corruptions, the place of the damned that needs to be enclosed and controlled, making 'city' and 'citizen' as politically opposed in the public imagination as they are etymologically linked.

This polarization of positive and negative images has its geography. Traditionally it registers as a division between secular and sacred space within the city. Later, the supposed virtues of the countryside and the small town were often contrasted with the evils of the city. When, for example, the rural army of reaction was assembled on the outskirts of Paris in 1871 poised to engage in the savage slaughter of some 30,000 communards, they were first persuaded that their mission was to reclaim the city from the forces of satan. When President Ford denied aid to New York City in 1975 in the midst of its fiscal crisis ('Ford to City: "Drop Dead!"' read the famous newspaper headline), the plaudits of virtuous and God-fearing small-town America were everywhere to be heard. In contemporary America, the image of the respectable God-fearing suburbs (predominantly white and middle class) plays against the inner city as a hell-hole where all the damned (with plenty of underclass racial coding thrown in) are properly confined. Imaginings of this sort take a terrible toll. When, for example, it was proposed to disperse some 200 families from the inner city of Baltimore to the suburbs as part of a 'Movement to Opportunity,' the suburbanites rose up in wrath to stop the program, using a language that sounded as if representatives of the devil were about to be released from their inner-city prison and let loose as a corrupting power in their midst. Religion doesn't always have to play this way of course. It also powers many an organization (like BUILD) that seeks to defend the poor, improve communities, and stabilize family life in the crumbling inner cities.

None of these imaginaries is innocent. Nor should we expect them to be. 'We make the house and the house makes us' is a saying that goes back to the Greeks. This was well understood by Robert Park (1967, 3), a founding figure of urban sociology, when he wrote:

[I]t is in the urban environment – in a world which man himself has made – that mankind first achieved an intellectual life and acquired those characteristics which most distinguish him from the lower animals and from primitive man. For the city and the urban environment represent man's most consistent and, on the whole, his most successful attempt to remake the world he lives in more after his heart's desire. But if the city is the world which man created, it is the world in which he is henceforth condemned to live. Thus, indirectly, and without any clear sense of the nature of his task, in making the city man has remade himself.

While we can reasonably aspire to intervene in that process of 'remaking ourselves' and perhaps even to acquire some 'clear sense of the nature of (our) task,' we cannot leap outside of the dialectic and imagine we are not embedded and limited by the institutional worlds and built environments we have already created. Yet we cannot evade the question of the imagination either for, as Marx (1976 edition, 283–4) observed (in a foundational statement that we will later examine in much greater depth): what distinguishes human labor and the worst of architects from the best of bees is that architects erect a structure in the imagination before realizing it in material form. When, therefore, we contemplate urban futures we must always do battle with a wide range of emotive and symbolic meanings that both inform and muddle our sense of 'the nature of our task.' As we collectively produce our cities, so we collectively produce ourselves. Projects concerning what we want our cities to be are, therefore, projects concerning human possibilities, who we want, or, perhaps even more pertinently, who we do not want to become. Every single one of us has something to think, say, and do about that. How our individual and collective imagination works is, therefore, crucial to defining the labor of urbanization. Critical reflection on our imaginaries entails, however, both confronting the hidden utopianism and resurrecting it in order to act as conscious architects of our fates rather than as 'helpless puppets' of the institutional and imaginative worlds we inhabit. If, as Unger (1987b, 8) puts it, we accept that 'society is made and imagined,' then we can also believe that it can be 'remade and reimagined.'

3 Utopianism as spatial play

Any project to revitalize utopianism needs to consider how and with what consequences it has worked as both a constructive and destructive force for change in our historical geography.

Consider Sir Thomas More's *Utopia*. More's aim, and this is characteristic, was social harmony and stability (in contrast to the chaotic state of affairs in England at that time). To this end, he excluded the potentially disruptive social forces of money, private property, wage labor, exploitation (the workday is six hours), internal (though not external) commodity exchange, capital accumulation, and the market process (though not a market place). The happy perfection of the social and moral order depends upon these exclusions. All of this is secured, as Lukerman and Porter (1976) point out, by way of a tightly organized spatial form (Plate 8.18). Utopia is an artificially created island which functions as an isolated, coherently organized, and largely closed-space economy (though closely monitored relations with the outside world are posited). The internal spatial ordering of the island strictly regulates a stabilized and unchanging social process. Put crudely, spatial form controls temporality, an imagined geography controls the possibility of social change and history.

Not all forms of temporality are erased. The time of 'eternal return', of recurrent ritual, is preserved. This cyclical time, as Gould (1988) remarks, expresses 'immanence, a set of principles so general that they exist outside of time and record a universal character, a common bond, among all of nature's rich particulars,' including, in this instance, all the inhabitants of Utopia. It is the dialectic of social process that is repressed. Time's arrow, 'the great principle of history,' is excluded in favor of perpetuating a happy stationary state. No future needs to be envisaged because the desired state is already achieved. In Bacon's *New Atlantis,* a utopian text written shortly after More's, the King decides that society has achieved such a state of perfection that no further social change is needed. In Bacon's case, technological change and new knowledges are not only deemed possible but actively sought. But their implantation is tightly regulated by the wise men of Salomon's House (an institution interpreted as a forerunner of the Royal Society). The effect is to progress towards the technological and learned perfection of an already perfected social order. More, by contrast, evokes nostalgia for a mythological past, a perfected golden age of small-town living, a stationary-state moral order and a hierarchical mode of social relating that is non-conflictual and harmonious. This nostalgic strain is characteristic of much utopian thinking, even that projected into the future and incorporating futuristic technologies. As we shall see, it has important consequences for how, if at all, such schemes get translated into material fact.

There are many ways to understand More's text and the numerous utopian schemas produced subsequently (such as those of Bacon and Campanella). I isolate here just one aspect: the relationship proposed between space and time, between geography and history. All these forms of Utopia can be characterized as 'Utopias of spatial form' since the temporality of the social process, the dialectics of social change – real history – are excluded, while social stability is assured by a fixed spatial form. Louis

Plate 8.18 Thomas More's Utopia: *an exercise in spatial play. Holbein's 'Frontispiece' to More's* Utopia *captures some of its spatial structure and its consequent spatial controls over the moral and political order.*

Marin (1984) considers More's Utopia as a species of 'spatial play.' More in effect selects one out of many possible spatial orderings as a way to represent and fix a particular moral order. This is not a unique thought. Robert Park (1967), for example, wrote a compelling essay in 1925 on the city as 'a spatial pattern and a moral order' and insisted upon an inner connection between the two. But what Marin opens up for us is the idea that the free play of the imagination, 'utopics as spatial play,' became, with More's initiative, a fertile means to explore and express a vast range of competing ideas about social relationships, moral orderings, political-economic systems, and the like.

The infinite array of possible spatial orderings holds out the prospect of an infinite array of possible social worlds. What is so impressive about subsequent utopian plans when taken together is their variety. Feminist utopias of the nineteenth century (Hayden, 1981) look different from those supposed to facilitate easier and healthier living for the working class and all sorts of anarchist, ecologically-sensitive, religious, and other alternatives define and secure their moral objectives by appealing to some specific spatial order (Plates 8.19, 8.20 and 8.21). The range of proposals – and of spatialities – testifies to the capacity of the human imagination to explore socio-spatial alternatives (see, e.g., Bloch, 1988; Kumar, 1987; 1991; Levitas, 1990; Sandercock, 1998). Marin's notion of 'spatial play'

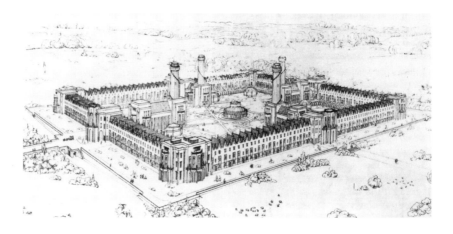

Plate 8.19 A design for Robert Owen's New Harmony. Robert Owen, one of the most prolific and fecund of utopian writers and activists in Britain in the first half of the nineteenth century, actually put some of this utopian schemes into practice. Stedman Whitewell proposed the above design for Owen's New Harmony Settlement in the United States.

Plate 8.20 Fourier's ideal city. Fourier drew for inspiration upon the layout of Versailles in his plan for a collectively organized communist industrial society dominated by communal production and communal living arrangements.

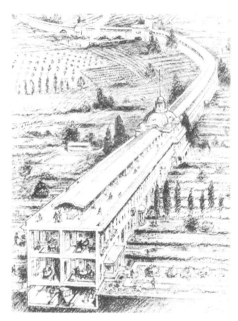

Plate 8.21 Edward Chambless: Road-town. The new systems of transportation led many designers to break with the traditional circular motif in favor of linear settlements oriented to major communication links. In this design two levels of dwellings with cooperative housekeeping arrangements spread throughout are underlain by a 'soundless' monorail and overlain with an extensive open promenade. This design, by Edward Chambless from 1910, sparked considerable interest in women's journals of the time.

neatly captures the free play of the imagination in utopian schemes. Reversion to this utopian mode appears to offer a way out of Unger's dilemma.

Alas matters are not so simple. Imaginative free play is inextricably bound to the existence of authority and restrictive forms of governance. What Foucault regards as 'a panoptican effect' through the creation of spatial systems of surveillance and control (polis = police) is also incorporated into utopian schemes. This dialectic between imaginative free play and authority and control throws up serious problems. The rejection, in recent times, of utopianism rests in part on an acute awareness of its inner connection to authoritarianism and totalitarianism (More's *Utopia* can easily be read this way). But rejection of utopianism on such grounds has also had the unfortunate effect of curbing the free play of the imagination in the search for alternatives. Confronting this relationship between spatial play and authoritarianism must, therefore, lie at the heart of any regenerative politics that attempts to resurrect utopian ideals. In pursuing this objective, it is useful to look at the history of how utopias have been materialized through political-economic practices: it is here that the dialectic of free play of the imagination and authoritarianism comes to life as a fundamental dilemma in human affairs.

4 Materializations of utopias of spatial form

All the great urban planners, engineers, and architects of the twentieth century set about their tasks by combining an intense imaginary of some alternative world (both physical and social) with a practical concern for engineering and re-engineering urban and regional spaces according to radically new designs. While some, such as Ebenezer Howard (Plate 8.22), Le Corbusier (Plate 8.23), and Frank Lloyd Wright (Plate 8.24) set up the imaginative context, a host of practitioners set about realizing those dreams in bricks and concrete, highways and tower blocks, cities and suburbs, building versions of the Villes Radieuse or Broadacre City (Plate 8.24), whole new towns, intimate scale communities, urban villages, or whatever. Even when critics of the authoritarianism and blandness of these realized utopian dreams attacked them, they usually did so by contrasting their preferred version of spatial play with the spatial orderings that others had achieved.

When, for example, Jane Jacobs (1961) launched her famous critique of modernist processes of city planning and urban renewal (damning as she did so Le Corbusier, the Charter of Athens, Robert Moses, and the great blight of dullness they and their acolytes had unleashed upon post-war cities), she in effect set up her own preferred version of spatial play by appeal to a nostalgic conception of an intimate and diverse ethnic neighborhood in which artisan forms of entrepreneurial activity and employment and interactive face-to-face forms of social relating predominated. Jacobs was in her own way every bit as utopian as the utopianism she attacked. She proposed to play with the space in a different and more intimate (scaled-down) way in order to achieve a different kind of moral purpose. Her version of spatial play contained its own authoritarianism hidden within the organic notion of neighborhood and community as a basis for social life. The apparatus of surveillance and control that she regarded as so benevolent, because it provided much-needed security, struck others, such as Sennett (1970), as oppressive and demeaning. And while she placed great emphasis upon social diversity, it was only a certain kind of controlled diversity that could really work in the happy way she envisaged. Pursuit of Jacobs's goals could easily justify all those 'intimately designed' gated communities and exclusionary communitarian movements that now so fragment cities across the United States.

This brings us to perhaps the most intriguing of Marin's categories: that of 'degenerate utopias.' The example that Marin used was Disneyland, a supposedly happy, harmonious, and non-conflictual space set aside from the 'real' world 'outside' in such a way as to soothe and mollify, to entertain, to invent history and to cultivate a nostalgia for some mythical past, to

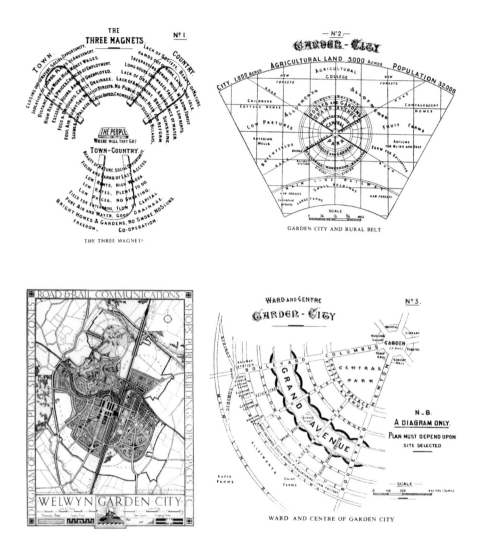

THE THREE MAGNETS

THE THREE MAGNETS

GARDEN CITY AND RURAL BELT

WARD AND CENTRE OF GARDEN CITY

Plate 8.22 Ebenezer Howard: from spatial ideals to new towns. *Ebenezer Howard, inspired by his reading of Edward Bellamy's utopian novel* Looking Backward, *set out to construct a whole new framework for urban living in his famous texts of 1898 and 1902. The 'new towns movement' that he sparked has arguably been one of the most influential strains of urban planning thought in the twentieth century.*

Plate 8.23 Le Corbusier's dream of the ideal city: theory and practice. Le Corbusier's *'Dream for Paris'* of the 1920s became the basis of the urban theory later incorporated in the very influential Charter of Athens. It appears largely realized in the achieved design for Stuyvesant Town, New York.

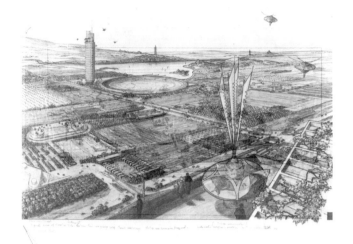

Plate 8.24 Frank Lloyd Wright's plan for Broadacre City. Concerned about the effects of the social collapse of the 1930s and influenced by the abundance of land in the American West and Midwest in particular, Frank Lloyd Wright proposed an alternative organization of space that permitted much greater degrees of personal independence while retaining communicative connections. The landscape he proposed bears a certain kind of corrupted resemblance to contemporary suburban sprawl (see Plate 8.4) which incorporates all the negatives of Wright's design without supporting any of its positive features.

perpetuate the fetish of commodity culture rather than to critique it. Disneyland eliminates the troubles of actual travel by assembling the rest of the world, properly sanitized and mythologized, into one place of pure fantasy containing multiple spatial orders. The dialectic is repressed and stability and harmony are secured through intense surveillance and control. Internal spatial ordering coupled with hierarchical forms of authority preclude conflict or deviation from a social norm. Disneyland offers a fantasy journey into a world of spatial play. And in its later incarnations, as at Epcot, it offers a futuristic utopia of technological purity and unsurpassed human power to control the world (Disney moved, as it were, from More to Bacon for his inspiration). All of this is degenerate, in Marin's view, because it offers no critique of the existing state of affairs on the outside. It merely perpetuates the fetish of commodity culture and technological wizardry in a pure, sanitized, and a-historical form. But, and this is where Marin's idea becomes problematic, Disneyland is an actual built environment and not an imagined place of the sort that More and Bacon produced. This immediately raises the question: can any utopianism of spatial form that gets materialized be anything other than 'degenerate' in the sense that Marin has in mind? Perhaps Utopia can never be realized without destroying itself. If so, then this profoundly affects how any utopianism of spatial form can function as a practical social force within political-economic life.

Generalizing from Marin, it can be argued that we are surrounded by a whole host of degenerate utopias of which Disneyland is but the most spectacular exemplar. When 'the malling of America' became the vogue, pioneers like James Rouse (Plates 8.8 and 8.12), who incidentally built his prototype mall in a Baltimore suburb and later returned to build the pavilions that anchor the Inner Harbor renewal, explicitly recognized that Disney had invented a formula for successful retailing. The construction of safe, secure, well-ordered, easily accessible, and above all pleasant, soothing, and non-conflictual environments for shopping was the key to commercial success. The shopping mall was conceived of as a fantasy world in which the commodity reigned supreme. And if homeless old folks started to regard it as a warm place to rest, youths found it a great place to socialize, and political agitators took to passing out their pamphlets, then the apparatus of surveillance and control (with hidden cameras and security agents) made sure nothing untoward happened (Plate 8.9).

As Benjamin (1969) remarked on the Parisian arcades of the nineteenth century, the whole environment seemed designed to induce nirvana rather than critical awareness. And many other cultural institutions – museums and heritage centers, arenas for spectacle, exhibitions, and festivals – seem to have as their aim the cultivation of nostalgia, the production of sanitized collective memories, the nurturing of uncritical aesthetic sensibilities, and the absorption of future possibilities into a non-conflictual arena that is eternally present. The continuous spectacles of commodity culture, including the commodification of the spectacle itself, play their part in fomenting political indifference. It is either a stupefied nirvana or a totally blasé attitude (the fount of all indifference) that is aimed at (Simmel [1971] long ago pointed to the blasé attitude as one of the responses to excessive stimuli in urban settings). The multiple degenerate utopias that now surround us – the shopping malls and the 'bourgeois' commercialized utopias of the suburbs being paradigmatic – do as much to signal the end of history as the collapse of the Berlin Wall ever did. They instantiate rather than critique the idea that 'there is no alternative,' save those given by the conjoining of technological fantasies, commodity culture, and endless capital accumulation (Plate 8.8).

James Rouse, incidentally, in one of those local ironies for which Baltimore is quietly famous, returned to the inner city after he retired to be an active participant, through his Enterprise Foundation, in a program of neighborhood revitalization in a community called Sandtown Winchester. There, his activities are memorialized (Plate 8.12) as he sought paternalistically to help rebuild a community undermined in part through the very processes of suburbanization and commercialization that he had in his business incarnation done so much to promote.

But how could it happen that the critical and oppositional force given in utopian schemes so easily degenerates in the course of materialization into compliance with the prevailing order? There are, I think, two basic answers to this question. Let me unpack them by a closer look at what is now held out as one of the leading candidates to transform our urban futures, the movement called 'the new urbanism.'

Duany (1997), one of its leading lights, 'feels strongly that urbanism, if not architecture, can affect society.' Getting the spatial play right, in the manner proposed by the new urbanism will, he argues, help rectify matters. His proposals evidence a nostalgia for small-town America, its solid sense of community, its institutions, its mixed land uses and high densities, and its ideologists (such as Raymond Unwin). Bring all this back in urban design and the quality of urban living and of social life will be immeasurably improved. This argument is buttressed by appeal to a long line of critical commentary (Kunstler, 1993; 1996) on the 'placelessness' and the lack of 'authenticity' in American cities (soulless sprawling suburbs, mindless edge cities, collapsing and fragmenting city cores fill in the pieces of this dispeptic view). The new urbanism does battle with such monstrous deformities (Katz, 1994). How to recuperate history, tradition, collective memory, and the sense of belonging and identity that goes with them becomes part of its holy grail. This movement does not, therefore, lack a critical utopian edge.

The new urbanism offers something positive as well as nostalgic. It does battle with conventional wisdoms entrenched in a wide range of institutions (developers, bankers, governments, transport interests, etc.). In the tradition of Mumford, it is willing to think about the region as a whole and to pursue a much more organic, holistic ideal of what cities and regions might be about. The postmodern penchant for fragmentation is rejected. It attempts intimate and integrated forms of development that by-pass the rather stultifying conception of the horizontally zoned and large-platted city. This liberates an interest in the street and civic architecture as arenas of sociality. It also permits new ways of thinking about the relation between work and living, and facilitates an ecological dimension to design that goes beyond superior environmental quality as a consumer good. It pays attention to the thorny problem of what to do with the profligate energy requirements of the automobile-based form of urbanization and suburbanization that has predominated in the United States since World War II. Some see it as a truly revolutionary force for urban change in the United States today.

But there are problems with materializing this utopian vision. The movement presumes that America is 'full of people who long to live in real communities, but who have only the dimmest idea of what that means in terms of physical design' (Kunstler, 1996). Community will rescue us from the deadening world of social dissolution, grab-it-yourself materialism, and individualized selfish market-oriented greed. But what kind of 'community' is understood here? Harking back to a mythological past of small-town America carries its own dangerous freight. The new urbanism connects to a facile contemporary attempt to transform large and teeming cities, so seemingly out of control, into an interlinked series of 'urban villages' where, it is believed, everyone can relate in a civil and urbane fashion to everyone else. In Britain, Prince Charles has led the way on this emotional charger towards 'the urban village' as the locus of urban regeneration. Leon Krier, an oft-quoted scion of the new urbanism, is one of his key architectural outriders. And the idea attracts, drawing support from marginalized ethnic populations, impoverished and embattled working-class populations left high and dry through deindustrialization, as well as from middle- and upper-class nostalgics who think of it as a civilized form of real-estate development encompassing sidewalk cafes, pedestrian precincts, and Laura Ashley shops.

The darker side of this communitarianism remains unstated. The spirit of community has long been held as an antidote to threats of social disorder, class war and revolutionary violence (More pioneered such thinking). Well-founded communities often exclude, define themselves against others, erect all sorts of keep-out signs (if not tangible walls), internalize surveillance, social controls, and repression. Community has often been a barrier to, rather than facilitator of, social change. The founding ideology of the new urbanism is both utopian and deeply fraught. In its practical materialization, the new urbanism builds an image of community and a rhetoric of place-based civic pride and consciousness for those who do not need it, while abandoning those that do to their 'underclass' fate. Most of the projects that have materialized are 'greenfield' developments for the affluent (including, of course, Prince Charles's own venture in the construction of Poundbury in Dorset, Plate 8.25). They help make the suburbs or the ex-urbs better places to live (Langdon, 1994). But they do little or nothing to help revitalize decaying urban cores. Scully (1994), a sceptical ally of the movement, doubts if the new urbanism can ever get to the crux of urban impoverishment and decay. In commenting on Seaside, that icon of the new urbanism, he notes that it has 'succeeded beyond any other work of architecture in our time . . . in creating an image of community, a symbol of human culture's place in nature's vastness' (the same is now being said, by the way, of Prince Charles's Poundbury). But, Scully continues:

> [O]ne cannot help but hope that the lessons of Seaside and of the other new towns now taking shape can be applied to the problem of housing for the

Plate 8.25 Poundbury, Dorset. *Prince Charles has led the way in a movement that calls for the construction of 'urban villages' as a solution to big city problems. He has put these ideals to work on one of his own estates close to Dorchester, and constructed a high density neighborhood appealing to the nostalgia of vernacular styles and small-town intimacies that were supposed to characterize a bygone era.*

poor. That is where community is most needed and where it has been most disastrously destroyed. Center city would truly have to be broken down into its intrinsic neighborhoods if this were to take place within it. Sadly, it would all have been much easier to do before Redevelopment, when the basic structure of neighborhoods was still there ... It is therefore a real question whether 'center city' as we know it can ever be shaped into the kind of place most Americans want to live in. (229)

The presumption here is that neighborhoods are in some sense 'intrinsic,' that the proper form of cities is some 'structure of neighborhoods,' that 'neighborhood' is equivalent to 'community' and 'community' is what most Americans want and need (whether they know it or not). It is further presumed that action at the scale defined by this new urbanism is effective and sufficient to solve problems that exist at all other scales. The nostalgic and spatially limited strain of the utopian dream resurfaces.

All of this happens because the 'new urbanism' must, if it is to be realized, embed its projects in a restrictive set of social processes. Duany (1997), for example, declares he has no interest in designing projects that will not get built. His concern for low-income populations is limited by a minimum price for new housing units in a place like Kentlands (Plate 8.26), not too far from Baltimore, of $150,000 (nearly ten times the median income in Baltimore). His interest in the suburbs arose simply because this is where most new projects can be built. Suburban growth, he argues, is

Plate 8.26 Utopian nostalgia: the commercialized new urbanism of Kentlands, Maryland. *Kentlands, designed by Andres Duany and Elizabeth Plater-Zyberk, is billed as a revolutionary approach to the new urbanism. Placed in the midst of a 'technoburb' (housing the National Bureau of Standards, IBM and several other high tech companies) it offers 'old-fashioned urban planning' with high densities, sidewalks and 'small town charm'. Close to Washington and not far from Baltimore, Kentlands is billed as a stellar example of the 'new urbanism' at work, offering accommodations to more than 5,000 on a totally planned 356 acre site.*

Single family homes begin at around $400,000, townhomes begin at $250,000 and provision for any low income population does not go below $150,000 for a condo unit. The high density design offers mainly an eclecticism of architectural styles and white picket fences that echo the strange world of David Lynch's 'Blue Velvet'. The cars (mainly gas-guzzling Sports Utilities Vehicles) are housed better than two-thirds of the world's population and a nod to ecological benevolence is given by the existence of a pond, the preservation of a few patches of woodland and naming the Elementary School after Rachel Carson. It is served by a standard suburban shopping center manned by workers who certainly cannot afford to live in the community they serve. While innovative relative to suburban sprawl, Kentlands sells small town nostalgia in a suburban setting to a very affluent clientele.

'the American way,' buried deep 'in our culture and our tradition' and while he objects strongly to the accusation that he is 'complicit' with power structures and that he panders to popular taste, he also insists that everything he does is designed to create spectacular projects that outperform all others on a commercial basis. This means 'faster permits, less cost, and faster sales.' His version of the new urbanism operates strictly within such parameters.

But who is at fault here? The designer, Duany, or the conditions of the social process that define the parameters of his projects? In practice, most realized Utopias of spatial form have been achieved through the agency of either the state or capital accumulation, with both acting in concert being the norm (this is as true in Singapore and Korea as it is in Britain, Sweden, France, or Australia). It is either that, or moving 'outside' of mainstream social processes (as seemed possible at least in the nineteenth century, with the United States being a favored target for utopian idealists such as Cabet, Robert Owen, and multiple religious movements). Those who took such an outsider path typically suffered a kind of meltdown of their principles, however, as they were absorbed within the mainstream of capital accumulation and the developmental state (something similar happened to the Israeli kibbutz).

The failure of realized utopias of spatial form can just as reasonably be attributed to the processes mobilized to materialize them as to failures of spatial form *per se*. This, as Tafuri (1976) so cogently argues, is what makes an architectural utopianism under present conditions such an utter impossibility. But there is a more fundamental contradiction at work here. Utopias of spatial form are typically meant to stabilize and control the processes that must be mobilized to build them. In the very act of realization, therefore, the historical process takes control of the spatial form that is supposed to control it. This contradiction requires further scrutiny.

5 On the utopianism of social process

If materialized utopias went wrong because of the social processes mobilized in their construction, then the focus switches to questions of process. Can we think of a utopianism of process rather than of spatial form? Idealized schemas of process abound. But we do not usually refer to them as 'utopian.' I want, along with a few select commentators like Polanyi, to break with that convention and consider the utopianism of temporal process alongside the utopianism of spatial form.

The use of the term 'utopian' may seem strange in this context because the word 'Utopia' is usually attached to some place that is no place as well as a happy place. The qualities of place (what might be called 'placefulness') are important and this means evocation of and close attention to spatial form as a container of social processes and as an expression of moral order. Idealized versions of social processes, in contrast, usually get expressed in purely temporal terms. They are literally bound to no place whatsoever and are typically specified outside of the constraints of spatiality altogether. The qualities of space and place are totally ignored.

We can identify a rich and complicated history of utopics as divergent temporal unfoldings. One obvious candidate is Hegel, whose guiding spirit is rendered material and concrete by a dialectics of transcendence (a dialectics that unfolds on the logic of 'both-and'). Things in themselves move history as they become things for themselves. The end state of history is, interestingly, expressed as a spatialized metaphor. The ethical or aesthetic state is the end point of the unfolding of the World Spirit. Marx sometimes followed this line of thinking though it was not the World Spirit but active class struggle that assumed the guiding role. As classes in themselves become classes for themselves, so history was moved onwards towards the perfected state of a post-revolutionary classless com-munistic society where even the state ultimately withered away. In both cases (and I obviously simplify) the ultimate stationary state as spatial form (which is unspecifiable in advance) is arrived at through a particular conception of historical process. Whereas More gives us the spatial form but not the process, Hegel and Marx give us their distinctive versions of the temporal process but not the ultimate spatial form.

There is, of course, plenty to protest in such placeless teleologies. Both William Blake and Kierkegaard, for example, insisted that the dialectic should be understood as 'either-or' rather than 'both-and.' The effect is to make history a succession of existential or political choices which have no necessary guiding logic or any clearly identifiable end state (Clark, 1991). Put another way (and this is a crucial point to which I will return), utopias of the social process have the habit of getting lost in the romanticism of endlessly open projects that never have to come to a point of closure (within space and place).

In detail, we find Marx in his political histories and later writings often drawn to a dialectics of 'either-or' rather than the 'both-and' of Hegelian transcendence. His hesitation in supporting the Paris Commune on the grounds that the time was not yet ripe and his sudden switch to support it up to the hilt had everything to do with his double sense of a dialectic that could be 'both-and' or 'either-or.' Marx clearly recognizes the potential consequences of either making a revolution or not in a given place and time and with this the teleology gives way to a much more contingent sense of historical unfolding, even if the motor of history still remains class struggle. As he wrote in his famous letter to Kugelmann on the subject:

World history would indeed be very easy to make, if the struggle were taken up only on condition of infallibly favourable chances. It would, on the other hand, be of a very mystical nature, if 'accidents' played no part.

(Marx and Lenin, 1940, 87)

Furthermore, the fact that it was in Paris that the Commune was occurring lent definite qualities (tangible strengths as well as weaknesses) to the movement while posing the question as to whether and how the revolutionary movement might move outwards from this epicenter to engulf the whole of France and even Europe. The distinction between the dialectic of an open-ended 'both-and' and the closure presupposed in 'either-or' is, as we shall see, no trivial matter.

In order to sustain his views Marx had to deconstruct a different and even then dominant utopianism of process that relied upon the rational activities of 'economic man' in a context of perfected markets. Since this has been by-far the most powerful utopianism of process throughout the history of capitalism we need to pay close attention to it. Adam Smith articulated the argument most precisely in *The Wealth of Nations*, first published in 1776. His reflections on the theory of moral sentiments – he was in the first instance a moral philosopher rather than an economist – led him to propose a utopianism of process in which individual desires, avarice, greed, drives, creativity, and the like could be mobilized through the hidden hand of the perfected market to the social benefit of all. From this Smith and the political economists derived a political program to eliminate state interventions and regulations (apart from those that secured free-market institutions) and curb monopoly power. *Laissez-faire*, free trade, and properly constituted markets became the mantras of the nineteenth-century political economists. Give free markets room to flourish, then all will be well with the world. And this, of course, is the ideology that has become so dominant in certain of the advanced capitalist countries (most notably the United States and Britain) these last twenty years. This is the system to which, we are again and again told, 'there is no alternative.'

Marx mounted a devastating attack upon this utopianism of process in *Capital*. In the second chapter he concedes the Smithian fiction of a perfected market. Then with a relentless and irrefutable logic he shows the inevitable consequences. An unregulated free-market capitalism, he proves, can survive 'only by sapping the original sources of all wealth – the soil and the laborer,' making the despoliation and degradation of the relation to nature just as important as the devaluation and debasement of the laborer. Furthermore:

[I]n proportion as capital accumulates, the situation of the worker, be his situation high or low, must grow worse ... Accumulation of wealth at one pole is, therefore, at the same time accumulation of misery, the torment of labour, slavery, ignorance, brutalization and moral degradation at the opposite pole ...   (1976 edition, 799)

Marx's brilliant deconstruction of free-market utopianism has largely been suppressed in recent times. Polanyi (writing in America during the Second World War with the Cold War clouds already on the horizon) understood Marx's point all too well and elaborated on it (without attribution) in the following terms:

> Our thesis is that the idea of a self-adjusting market implied a stark utopia. Such an institution could not exist for any length of time without annihilating the human and natural substance of society; it would have physically destroyed man and transformed his surroundings into a wild- erness. Inevitably, society took measures to protect itself, but whatever measures it took impaired the self-regulation of the market, disorganized industrial life, and thus endangered society in yet another way. It was this dilemma which forced the development of the market system into a definite groove and finally disrupted the social organization based upon it.
>
>   (Polanyi, 1957, 3–4)

But the rise of neoliberalism as a dominant ideology in the Thatcher– Reagan years (and its export around the world through a mix of persuasion and economic force) swept such objections aside. The free-market jug- gernaut, with its mantras of private and personal responsibility and initiative, deregulation, privatization, liberalization of markets, free trade, downsizing of government, draconian cut-backs in the welfare state and its protections, has rolled on and on. For more than twenty years now we have been battered and cajoled at almost every turn into accepting the utopian- ism of process of which Smith dreamed as the solution to all our ills. We have also witnessed an all-out assault on those institutions – trade unions and government in particular – that might stand in the way of such a project. Margaret Thatcher proclaimed that there is no such thing as society, only individuals and their families, and set about dismantling all those institutions – from trade unions to local governments – that might stand in the way of her utopian vision. With the fall of the Berlin Wall, Fukuyama put a Hegelian gloss on all of this. We are now at the end of history. Capitalism and the free market are triumphant world wide. The end of history is here (a sad thought if Baltimore is anything to go by).

It may seem strange to view the likes of Thatcher and Gingrich as Heg- elians, but the free-market triumphalism they espoused was nothing other than Smithian utopianism of process attached to a very Hegelian kind of teleology ('progress is inevitable and there is no alternative'). In many respects, as Frankel (1987) points out, the most effective utopians in recent times have been those of a right-wing persuasion and they have primarily espoused a utopianism of process rather than a utopianism of spatial form. The odd thing, however, has been the failure to attach the negative epithets of 'utopian' and 'teleological' to this right-wing assault upon the social order. Only recently has a main-stream thinker like John Gray sought to rehabilitate Polanyi and attack the inherent destructiveness of free-market utopianism. The precipitous fall from power and influence of both Thatcher and Gingrich testifies to their vulnerability on precisely such a count.

For the consequences of their utopianism when rendered actual are close to those that Marx's deconstruction depicts. Income inequalities have risen rapidly in all those countries that have given themselves over most energetically to the utopianism of the market (see Chapter 3). This polarization in income and wealth also has its geographical forms of expression: spiraling inequalities between regions as well as escalating contrasts between affluent neighborhoods and impoverished shanty towns or, in the case of the United States, between impoverished inner cities and affluent and exclusionary suburbs. Accelerating uneven geographical development, the undermining of all forms of social cohesion and state powers, the destruction of whole cultures and of those 'structures of feeling' that give a solid grounding to daily life, and, perhaps most problematic of all, the degradation of large swathes of the environment to the degree that much of the earth's surface becomes uninhabitable, are some of the effects that Gray (1998, 207) correctly depicts. 'As a result,' he writes, 'we stand on the brink not of the era of plenty that free-marketeers project, but a tragic epoch, in which anarchic market forces and shrinking natural resources drag sovereign states into ever more dangerous rivalries.' These are, I insist, exactly the forces that are at work in the degradation of Baltimore. So why such tragic outcomes to such a supposedly benevolent process?

The answer broadly lies in what happens when the utopianism of process comes geographically to earth. For any materialization of free-market utopianism requires that the process come to ground someplace, that it construct some sort of space within which it can function. How it gets framed spatially and how it produces space become critical facets of its tangible realization. Much of my own work these last twenty years (Harvey, 1982; 1989) has been about trying to track exactly such a process, to understand how capital builds a geographical landscape in its own image at a certain point in time only to have to destroy it later in order to accommodate its own dynamic of endless capital accumulation, strong technological change, and fierce forms of class struggle. The history of creative destruction and of uneven geographical development in the bourgeois era is simply stunning. Much of the extraordinary transformation of the earth's surface these last 200 years reflects precisely the putting into practice of the free-market utopianism of process and its restless and perpetual reorganizations of spatial forms.

But the conditions and manner of this spatial materialization have all manner of consequences. As free-market capital accumulation plays across a variegated geographical terrain of resource endowments, cultural histories, communications possibilities, labor quantities and qualities (a geographical terrain that is increasingly a differentiated product of capital investments in infrastructures, 'human capital,' and built environments), so it produces an intensification of uneven geographical development in standards of living and life prospects. Rich regions grow richer leaving poor regions ever poorer (Baltimore provides a dramatic example of such uneven geographical development at the metropolitan scale). Circular and cumulative causation embedded within the utopianism of the market process produces increasing geographical differentiations in wealth and power, rather than gradual progress towards homogeneity and equality. There is, the saying goes, 'nothing more unequal than the equal treatment of unequals' and it is precisely on this point that the egalitarianism and the democratization implicit in freely functioning markets produces more rather than less inequality in the long run.

Community and/or state power has led the way in trying to counteract some of the more egregious consequences of free-market utopianism (spiraling income inequalities, uneven geographical developments, externality effects on the environment, and the like). But there is a deep paradox that lurks in this phenomenon. The free market, if it is to work, requires a bundle of institutional arrangements and rules that can be guaranteed only by something akin to state power. The freedom of the market has to be guaranteed by law, authority, force, and, *in extremis*, by violence. Since state power is usually understood in terms of the monopoly of the forces of violence, the free market requires the state or cognate institutions if it is to work. Free markets, in short, do not just happen. Nor are they antagonistic to state power in general though they can, of course, be antagonistic to certain ways in which state power might be used to regulate them.

The play of free-market utopianism can be assured only if, as Marx and Engels pointed out in the *Communist Manifesto*, the state (and we must now include the local state in this determination) becomes 'the executive committee of the bourgeoisie.' Decolonization after 1945, and the subsequent internationalization and liberalization of global markets, has brought the whole world much closer to that norm, though the uneven pace at which this has occurred (a product of political and social struggles in which resort to traditional solidarities and traditions has played an important role) has affected how the utopianism of process has been materialized in different places and times. Geopolitical struggles between places and, even more destructively, between states or alliances of states are integral to the problem.

The upshot of this argument is that the purity of any utopianism of process inevitably gets upset by its manner of spatialization. In exactly the same way that materializations of spatial utopias run afoul of the particularities of the temporal process mobilized to produce them, so the utopianism of process runs afoul of the spatial framings and the particularities of place construction necessary to its materialization. Adam Smith, curiously, saw the problem. It was hard, he wrote, to foresee the mix of benefits and misfortunes that extension of market exchange might confer on different parts of the world. But, he hoped:

> [B]y uniting, in some measure, the most distant parts of the world, by enabling them to relieve one another's wants, to increase one another's enjoyments, and to encourage one another's industry, their general tendency would seem to be beneficial. To the natives, however, both of the East and West Indies, all the commercial benefits which can have resulted from these events have been sunk and lost in the dreadful misfortunes which they have occasioned. These misfortunes, however, seem to have arisen rather from accident than from any thing in the nature of those events themselves. At the particular time when these discoveries were made, the superiority of force happened to be so great on the side of the Europeans, that they were enabled to commit with impunity every sort of injustice in those remote countries. Hereafter, perhaps, the natives of those countries may grow stronger, or those of Europe may grow weaker, and the inhabitants of all the different quarters of the world may arrive at that equality of courage and force which, by inspiring mutual fear, can alone overawe the injustice of independent nations into some sort of respect for the rights of one another. But nothing seems more likely to establish this equality of force than that mutual communication of knowledge and of all sorts of improvements which an extensive commerce from all countries to all countries naturally, or rather necessarily, carries along with it.
>
>   (Adam Smith, cited in Arrighi, 1994, 19)

The effects were by no means as accidental or as transitory as Smith's utopian vision supposed. Nor could Smith's standard response, readily to be found in writings now as then, that the problem of inequality arises because the perfection of the market has not yet been fully realized, carry weight and credibility after two centuries of hard experience.

6 Grounding social processes in spatial forms

Utopias of spatial form get perverted from their noble objectives by having to compromise with the social processes they are meant to control. We now see also that materialized utopias of the social process have to negotiate with spatiality and the geography of place and in so doing they also lose their ideal character, producing results that are in many instances exactly the opposite of those intended (e.g. increasing authoritarianism and inequalities rather than greater democracy and equality). Let us look more closely at exactly how such an inversion occurs in the case of idealized social processes.

There are two fundamental points at which a 'negotiation of spatiality' must occur when any utopianism of the social process gets materialized. Consideration of them illustrates how and why the effects that Smith lamented cannot be construed in any way as accidental or transitory. Free markets rely, as we have seen, upon state power. The development of free markets depends crucially upon the extension as well as the intensification of specific forms of state power. Contrary to popular belief, market processes do not lead to a 'hollowing out' of the state. They entail a deepening of the state's grasp over certain facets of the social process even as it is driven away from performing some of its other more traditional and populist functions. Furthermore, to the degree that the state itself requires legitimacy to perform its role most effectively, populist, nationalist, and imperialist sentiments must be mobilized in its support, turning the extension of the free market into a political or even more markedly geopolitical crusade. The British were pushers of market processes across the world in the nineteenth century using gunboat diplomacy, imperial conquest, and a whole range of notions about racial superiority, 'the white man's burden,' and convictions about their 'civilizing mission' in their drive to open up the world to trade. Americans sought a new world order of a free market coupled with a supposedly 'democratic' capitalism after 1945 using all the means of persuasion and violence at their command. In the last twenty years globalization and freedom of trade have become a crusading theme in American foreign policy, again indicating that hegemonic state power is essential to freely functioning markets. On a more local level, Margaret Thatcher could materialize her free-market philosophy within Britain only by ruthless use of state powers (e.g. police violence to quell strikes, strict supervision of university research) and by appeal to nationalist sentiments (the latter creating a paradox that led to her downfall as she refused the political integration with Europe that freedom of the market truly and logically demanded).

This signals a fundamental contradiction. The preservation and extension of state power is crucial to the functioning of free markets. If free markets, as is their wont, undermine state powers, then they destroy the conditions of their own functioning. Conversely, if state power is vital to the functioning of markets, then the preservation of that power requires the perversion of freely functioning markets. This is, as Polanyi so clearly outlines, the central contradiction that lies at the heart of neoliberal political economy. It explains why so much of the developmental pattern in a city like Baltimore is justified by appeal to the rhetoric of free-market competition when it in practice relies on state subsidy and monopolization. It also explains why the great eras of globalization and freer international trade have been those where a single power (such as Britain in the late nineteenth century or the United States after 1945) was in a position to guarantee the political, institutional, and military conditions for market freedoms to prevail.

A surface veneer of competitive capitalism therefore depends on a deep substratum of coerced cooperations and collaborations to ensure a framework for the free market and open trade.

The second fundamental point of negotiation of free-market utopianism with spatiality rests more directly with the construction of physical built environments as resource complexes upon which commercial activity can build. In its crassest forms this permits the formulation of the kind of commercialized utopianism that had someone like Margaret Thatcher set up Urban Development Corporations to revitalize urban zones (the London Docklands being the great example). But investment in infrastructures always generates geographical biases and uneven geographical developments that then also draw in yet more development as part of the synergism that inevitably arises as free-market activities engage in the production of space. The selling of a place, like Baltimore, then itself becomes part of the art of utopian presentation. And it is at this point that rhetorical flourishes drawn from utopias of spatial form get combined with rhetorical flourishes concerning a utopianism of process to produce commercialized and degenerate utopian forms all round us.

The outcome of such contradictions is a deepening rather than a lessening of uneven geographical development in both its political and economic dimensions. The extension of all manner of systems of state domination reduces whole zones of the world and various strata in the population living therein to conditions akin to servitude. And the concentration of mainly public resources in space produces spiraling geographical inequalities at all scales. And this all in the interests of preserving the political economic sources of state power that guarantee the functioning of free markets. The paradoxes and contradictions are everywhere self-evident. Yet the utopian rhetoric of freedom, liberty, and markets conceals so effectively that we often find it difficult to articulate the pattern of underlying coerced collaborations that otherwise stares us so blatantly in the face.

CHAPTER 9

Dialectical utopianism

1 Towards a spatiotemporal utopianism

Given the defects and difficulties of utopias of both spatial form and social process, the most obvious alternative (other than total abandonment of any pretense at utopianism whatsoever) is to build a utopianism that is explicitly spatiotemporal. It is many years now since Einstein taught us that space and time cannot meaningfully be separated. There are more than a few hints within the social sciences that the separation of space from time, though sometimes useful, can often be misleading (see Harvey, 1996, Part III). And if space and time are viewed as social constructs (implying the rejection of the absolute theories of space and time attributable to Newton and Descartes), then the production of space and time must be incorporated into utopian thought. The search is on, therefore, for what I shall call 'dialectical utopianism.'

The lessons to be learned from the separate histories of utopianisms of spatial form and temporal process must not, however, be abandoned. Indeed, there are even further insights to be had from a closer analysis of them. From the former, the idea of imaginative spatial play to achieve specific social and moral goals can be converted into the idea of potentially endlessly open experimentation with the possibilities of spatial forms. This permits the exploration of a wide range of human potentialities (different modes of collective living, of gender relations, of production-consumption styles, in the relation to nature, etc.). This is, for example, how Lefebvre (1991) sets up his conception of the production of space. He sees it as a privileged means to explore alternative and emancipatory strategies.

But Lefebvre is resolutely antagonistic to the traditional utopianisms of spatial form precisely because of their closed authoritarianism. He fashions a devastating critique of Cartesian conceptions, of the political absolutism that flows from absolute conceptions of space, of the oppressions visited upon the world by a rationalized, bureaucratized, technocratically, and capitalistically-defined spatiality. For him, the production of space must always remain as an endlessly open possibility. The effect, unfortunately, is to leave the actual spaces of any alternative frustratingly undefined. Lefebvre refuses specific recommendations (though there are some nostalgic hints that they got it right in Renaissance Tuscany). He refuses to confront the underlying problem: that to materialize a space is to engage with closure (however temporary) which is an authoritarian act. The history of all realized utopias points to this issue of closure as both fundamental and unavoidable, even if disillusionment through foreclosure is the inevitable consequence. If, therefore, alternatives are to be realized, the problem of closure (and the authority it presupposes) cannot endlessly be evaded. To do so is to embrace an agonistic romanticism of perpetually unfulfilled longing and desire. And this is, in the end, where Lefebvre leaves us.

Foucault sought to extricate himself from this same difficulty by a different path. In *The Order of Things*, first published in 1966, he coined the term 'heterotopia' to describe the incongruity, the 'enigmatic multiplicity' and the fundamental disorder of which language was itself so capable:

> Utopias afford consolation: although they have no real locality, there is nevertheless a fantastic untroubled region in which they are able to unfold; they open up cities with vast avenues, superbly planted gardens, countries where life is easy, even though the road to them is chimerical. Heterotopias are disturbing, probably because they secretly undermine language ... Utopias permit fables and discourse: they run with the very grain of language ... (heterotopias) desiccate speech, stop words in their tracks, contest the very possibility of grammar at its source; they dissolve our myths and sterilize the lyricism of our sentences.

In *The Order of Things*, 'heterotopia' is considered solely in relation to discourse and language. Foucault later sought to give the term a material referent. In a lecture given in 1967, in an attempt to shape a dialogue with architects and theorists of spatial form, Foucault (1986) appealed once again to the concept of 'heterotopia.' The lecture was never revised for publication (though he did permit its publication shortly before he died in 1984). Extracted by his acolytes as a hidden gem within his extensive *oeuvre*, it then became one means (particularly important within the canon of postmodernism) whereby the problem of Utopia could be resurrected and simultaneously evaded. The theme of 'escape' underwrites Foucault's essay. ('The ship is the heterotopia par excellence,' he wrote. 'In civilizations without boats, dreams dry up, espionage takes the place of adventure, and the police take the place of pirates' [1986, 27].) The concept allows Foucault to escape from the 'no place' that is a 'placeful' utopia (a theme that was animating much of the movement of 1968 in France) and come to earth in particular places of actual practices. But he also uses it to escape the world of norms and structures that imprison the human imagination (including, incidentally, his own anti-humanism) and, through a study of the history of space and an understanding of their heterogeneity, identify spaces in which difference, alterity, and 'the other' might flourish or (as with architects) actually be constructed. Hetherington (1997) summarizes this concept of heterotopia:

> as spaces of alternate ordering. Heterotopia organize a bit of the social world in a way different to that which surrounds them. That alternate ordering marks them out as Other and allows them to be seen as an example of an alternative way of doing things ... Heterotopia, therefore, reveal the process of social ordering to be just that, a process rather than a thing.

The formulation is surficially attractive. It allows us to think of the multiple utopian schemas (spatial plays) that have come down to us in materialized forms as not mutually exclusive. It encourages the idea of a simultaneity of spatial plays that highlights choice, diversity, and difference. It enables us to look upon the multiple forms of deviant and transgressive behaviors and politics that occur in urban spaces (Foucault interestingly includes in his list of heterotopic spaces such places as cemeteries, colonies, brothels, and prisons) as valid and potentially meaningful reassertions to some kind of right to shape parts of the city in a different image. It forces us to recognize how important it is to have spaces (the jazz club, the dance hall, the communal garden) within which life is experienced differently. There are, Foucault assures us, abundant spaces in which 'otherness,' alterity, and, hence, alternatives might be explored not as mere figments of the imagination but through contact with social processes that already exist. It is within these spaces that alternatives can take shape and from these spaces that a critique of existing norms and processes can most effectively be mounted. The history of such spaces, he asserts (drawing heavily on the work of Bachelard), shows us how and in what ways spatial forms might connect to radically different social processes and so disrupt the homogeneity to which society (and by extension its utopian antidotes) typically clings. He evidently hoped for that effect, earlier described in *The Order of Things*, of a 'disorder in which fragments of a large number of possible orders glitter separately in the dimension, without law or geometry, of the *heteroclite*' (Foucault, 1973, XVII).

Unfortunately, the concept cannot so easily escape the freight of utopias more generally (perhaps this is why Foucault refused to elaborate on the concept and even tacitly reneged on it in his *Discipline and Punish*). It presumes that connections to the dominant social order are or can be severed, attenuated or, as in the prison, totally inverted. The presumption is that power/knowledge is or can be dispersed and fragmented into spaces of difference. It presumes that whatever happens in such spaces of 'Otherness' is of interest and even in some sense 'acceptable,' or 'appropriate.' The cemetery and the concentration camp, the factory, the shopping malls and Disneylands, Jonestown, the militia camps, the open plan office, New Harmony, 'privatopia,' and 'ecotopia' are all sites of alternative ways of doing things and therefore in some sense 'heterotopic.' What appears at first sight as so open by virtue of its multiplicity suddenly appears either as banal (an eclectic mess of heterogeneous and different spaces within which anything 'different' – however defined – might go on) or as a more sinister fragmentation of spaces that are closed, exclusionary, and even threatening within a more comprehensive dialectics of historical and geographical transformation. The concept of 'heterotopia' has the virtue of insisting upon a better understanding of the heterogeneity of space but it gives no clue as to what a more spatiotemporal utopianism might look like. Foucault challenges and helps destabilize (particularly in the realm of discourse) but provides no clue as to how any kind of alternative might be constructed.

Consider the matter now from the standpoint of process-oriented utopias. The supposedly endlessly open and benevolent qualities of some utopian social process, like market exchange, have to crystallize into a spatially-ordered and institutionalized material world somewhere and somehow. Social, institutional, and material structures (walls, highways, territorial subdivisions, institutions of governance, social inequalities) are either made or not made. The dialectic of either-or is omnipresent. Once such structures are built they are often hard to change (nuclear power stations commit us for thousands of years and institutions of law gather more and more weight of precedent as time goes on). Struggle as we might to create flexible landscapes and institutions, the fixity of structures tends to increase with time making the conditions of change more rather than less sclerotic. A total reorganization of materialized organizational forms like New York City or Los Angeles is much harder to envisage let alone accomplish now than a century ago. Free-flowing processes become instantiated in structures, in institutional, social, cultural, and physical realities that acquire a relative permanence, fixity, and immovability. Materialized Utopias of process cannot escape the question of closure or the encrusted accumulations of traditions, institutional inertias, and the like, which they themselves produce. The more free-market utopianism converges on the inequalities and unfreedoms of actually existing capitalism, the harder it becomes to change or even maintain its own trajectory.

Any contemporary struggle to envision a reconstruction of the social process has to confront the problem of how to overthrow the structures (both physical and institutional) that the free market has itself produced as relatively permanent features of our world. Though daunting, that task is not impossible. The revolutionary agenda of neoliberalism has accomplished a lot in the way of physical and institutional change these last twenty years (consider the dual impact of deindustrialization and the diminution of trade union powers in Britain and the United States, for example). So why, then, can we not envision equally dramatic changes (though pointing in a different direction) as we seek for alternatives?

It is at this point that it is useful to consider the works of Roberto Unger who, like Lefebvre, is deeply committed to the exploration of liberatory alternatives, but equally anxious to avoid the errors of traditional utopian formulations. Unger focuses on social processes and institutional/personal transformations. His critique of existing institutions and behaviors is strong and powerful, as might be expected. So how does he go about the task of envisioning alternatives? Unger avoids utopianism by insisting that alternatives should emerge out of critical and practical engagements with the institutions, personal behaviors, and practices that now exist (most directly, the arrangements arrived at through the Western versions of democracy that have emerged out of the long history of capitalism). He is, therefore, only interested in 'the next step in a trajectory' rather than in some universal principles of transformation or the description of some millenarian vision. Yet visionary thought and imaginative struggle are critical to this endeavor:

> Our thinking about ideals becomes visionary or external to the extent that it holds up a picture, however partial or fragmentary, of a radically altered scheme of social life and appeals to justifications that do not stick close to familiar and established models of human association. The visionary is the person who claims not to be bound by the limits of the tradition he or his interlocutors are in ... Notice that visionary thought is not inherently millenarian, perfectionist, or utopian (in the vulgar sense of the term). It need not and does not ordinarily present the picture of a perfected society. But it does require that we be conscious of redrawing the map of possible and desirable forms of human association, of inventing new models of human association and designing new practical arrangements to embody them.   (Unger, 1987b, 359–60)

At the heart of Unger's work lies a simple but powerful dialectical conception. Only by changing our institutional world can we change ourselves at the same time, as it is only through the desire to change ourselves that institutional change can occur. 'The more the technical and social divisions of labor present themselves in everyday life as a rigid grid of functional allocations,' he writes, 'the more they deserve to be smashed up at the microlevel of cultural-revolutionary defiance and incongruity as well as at the macrolevel of institutional innovation' (564). The objective, however, is to build a more radically empowered and empowering system of democratic governance that can be both liberatory and transformative. To this end, Unger envisions three key varieties of empowerment. The first opens up social life to practical experimentation, the second strengthens our 'self-conscious mastery over the institutional and imaginative frameworks of our social experience,' while the third helps 'cleanse group life of some of its capacity to entangle people in relations of dependence and domination and to turn them into the faceless representatives of predetermined roles' (363–4).

To this end Unger envisages three main spheres for institutional reconstruction: the constitution of government, the organization of the economy, and the system of rights. He struggles to avoid the romanticism of endlessly open possibilities by exploring a variety of proposals in these domains which will, he hopes, have the effect of both galvanizing the sense of possible alternatives while setting in motion personal transformations that soften the contrast between the 'structure preserving routine' of daily life in which we are all embedded and the 'structure transforming conflicts' that are usually manifest as revolutionary and destructive violence. In this way he hopes 'to free sociability from its script and to make us available to one another more as the originals we all know ourselves to be and less as the placeholders in a system of group contrasts' (563–4).

I consider Unger at some length here (though no way suffolently to do justice to an enormously rich and complicated three-volume study) because it seems to me that he goes far in paying attention to how visionary ideas might be materialized. His analysis is very much of this world. He now works with the Brazilian Workers Party on the constitutional and legal aspects of progressive political action. And in cities like Porto Alegre where the Workers Party has held political control for several years, some highly innovative means have been found to enhance popular empowerment and democratic forms of governance, many of which bear the marks of the sort of thinking that Unger represents (see Abers, 1998, for an account of the Porto Alegre experiments). Translated into a situation like that of Baltimore, these experiments could be very helpful indeed. We have much to learn from them.

Unger, however, has no particular spatial model for social ordering in mind – his whole presentation abstracts from spatial considerations throughout. I do not consider this fatal to his argument since it would not be hard to extend his method of critiquing and re-envisioning social institutions to the plane of spatial forms, thus converting his arguments into a more deliberatively spatiotemporal dynamic of progressive demo-cratization and empowerment (of the sort that is indeed being constructed on the ground in Porto Alegre).

Yet there are some serious difficulties with Unger's approach. He notes, for example, what he terms 'an astonishing gap between the alleged interest in alternatives and the lack of any tangible sign that this interest is real.' And he likewise notes how 'the illusions of deep logic social theory' (primarily represented by Marxism and structuralism) 'and the faith in the spontaneous creative powers of revolutionary action have disarmed the constructive political imagination of the left.' He then writes:

> The few who try to work out alternatives more considered than those found in the party platforms of the mainstream of leftist literature are quickly dismissed as utopian dreamers or reformist tinkerers: utopians if their proposals depart greatly from the established arrangements, tinkerers if they make modest proposals of change. Nothing worth fighting for seems practicable, and the changes that can be readily imagined often hardly seem to deserve the sacrifice of programmatic campaigns whose time chart so often disrespects the dimensions of an individual lifetime. If all of this were not enough, the would-be program writer still has a final surprise in store for him. He will be accused – sometimes by the very people who told him a moment before they wanted alternatives – of dogmatically anticipating the future and trying to steal a march on unpredictable circumstance, as if there were no force to Montaigne's warning that 'no wind helps him who does not know to what port he sails'.   (1987a, 443)

This is perceptive stuff. The effect, however, is to create a hesitation on everyone's part (including Unger's) in ever identifying to what port he or she might wish to sail. And it is at that point that Unger refuses closure around any one particular set of institutional arrangements or modes of social relating. Like Lefebvre, he wants to keep choices endlessly open. The harsh 'either/or' of the dialectic is evaded in favor of the softer (one of Unger's favored words) and more comforting politics of Hegelian trans-cendence. The anti-authoritarianism of liberatory political thought here reaches some sort of limit. There is a failure to recognize that the materialization of anything requires, at least for a time, closure around a particular set of institutional arrangements and a particular spatial form and that the act of closure is in itself a material statement that carries its own authority in human affairs. What the abandonment of all talk of Utopia on the left has done is to leave the question of valid and legitimate authority in abeyance (or, more exactly, to leave it to the moralisms of the conservatives – both of the neoliberal and religious variety). It has left the concept of Utopia, as Marin observes, as a pure signifier without any meaningful referent in the material world. And for many contemporary theorists – Unger among them – that is where the concept can and should remain: as a pure signifier of hope destined never to acquire a material referent. But the problem is that without a vision of Utopia there is no way to define that port to which we might want to sail.

2 Utopian dynamics

It would be wrong to depict the theoretical set called 'spatiotemporal utopianism' as totally empty. Indeed, there are many ways in which it might be filled. To begin with, the evolutionary style of argumentation adopted by Geddes and Mumford in their respective approaches to the city and region promotes a view of human activity in which the production of spaces at a changing scale is expressive of equally compelling and often conflictual economic, technological, and cultural needs (the city is to be construed, as Mumford put it, above all as 'a work of art'). Mumford's writings, however idiosyncratic and defective, are infused with a certain kind of spatiotemporal utopianism (although with a good deal of dystopian feeling with respect to some of the catastrophic choices made in more recent times).

But even more important, the recent evolution within the utopian genre of writing itself from that of a political tract thinly disguised within an often rather boring tale (as in the case of More's *Utopia* or even Butler's *Erewhon*) into the full-blown drama of the (sometimes epic) novel signals an important transition in sentiments and techniques. The novel, as an exploration of possible worlds (see Ronen, 1994) has now become the primary site for the exploration of utopian sentiments and sensibilities. Earlier hints of such a turn can be found in the works like Hesse's *Magister Ludi*, H. G. Wells, Aldous Huxley's *Brave New World*, and Ayn Rand's *The Fountainhead* (with plenty of dystopian fiction such as Orwell's *1984* being thrown in). But in recent times the tactic has become explicit in the writings of Ursula Le Guin, Doris Lessing, Marge Piercy, and many others (and it is instructive to note how many writers now working in this genre are women – as Levitas, 1990; 1993, points out).

Such novels typically recognize that societies and spatialities are shaped by continuous processes of struggle. The novel form lends itself, if need be, to a much stronger sense of spatiotemporal dynamics. The static and finally achieved spatial/institutional forms of classical utopias are jettisoned as either unachievable or, if achieved, unstable and still in transition to something else yet to be defined. Consider, for example, Kim Stanley Robinson's trilogy on the settlement of Mars. A voyage of exploration followed by active colonization of a distant planet empty of people (though not of distinctive qualities), it re-writes the historical geography of colonization as a long-term saga of transformations of environment and of socio-spatial forms on Mars. The saga is marked by struggles against the Martian environment as well as within the colonizing process itself – struggles that pit the first hundred settlers against later arrivals, the authorities on earth with the nascent society on Mars, and, above all, struggles among the colonizers themselves on a wide range of questions (such as forms of political representation), but most particularly the issue of the moral and political right to 'terraform' the Martian environment. The formation of society on Mars entails the production of a world that is continuously evolving new forms out of itself, but not in an arbitrary way. Each decision point marks an 'either-or' and who wins the battle (physically or ideologically) changes the trajectory of development without, however, necessarily carrying all of the opposition with it. A residue of power and argument is always left behind often to be resurrected later on as an alternative subversive force. The reader is not, therefore, introduced to a stable world already made and discovered, but is taken through the dialectics of making a new socio-ecological world.

I am not holding up Robinson's tale as some exemplar of how a spatio-temporal utopianism should be set down, though it does illustrate how a cultural form can be used to articulate an alternative spatiotemporal dynamics. There are, moreover, several dangers in relying solely upon novels as a source of inspiration. The displacement of utopianism to 'pure' literature (or art), for example, may mean that we fail to extract the political messages that come through so loud and clear from a political tract like More's *Utopia* or Bellamy's *Looking Backward*. It is hard to imagine Robinson's work inspiring a whole political movement of the sort that arose at the end of the nineteenth century in response to Bellamy's intervention. Worse still, artistic license easily glosses over the real difficulties of transformative action. As Levitas (1993, 265) observes:

> The main reason why it has become so difficult to locate utopia in a future credibly linked to the present by a feasible transformation is that our images of the present do not identify agencies and processes of change. The result is that utopia moves further into the realms of fantasy. Although this has the advantage of liberating the imagination from the constraint of what it is possible to imagine as possible – and encouraging utopia to demand the impossible – it has the disadvantage of severing utopia from the process of social change and severing social change from the stimulus of competing images of utopia.

We are, as it were, back to Unger's distinction between dreams that seem unrealizable and prospects that hardly seem to matter.

On this point, however, one reading of Robinson's trilogy provides a further crucial insight. His utopian tale carries within it innumerable cross-references to the actual historical geography of imperial conquest and colonial and neocolonial activity as promoted throughout the long history of capitalism. The historical and geographical referents of his tale are highly significant. While it is grounded in relation to the tangible qualities of the Martian environment (with all of its distinctive physical hazards) and appeals to futuristic technologies, it inevitably invokes the whole problematic of mastery of nature that has been so central to Western ways of thought since Francis Bacon and Descartes. The struggle to stay alive in the face of a hostile Martian environment is nothing short of heroic. Furthermore, the embeddedness of the colonization process on Mars in the power relations, the ideological debates, and the technological political-economy of multinational and globalized capitalism is explicitly acknowledged. The colonization of Mars is a struggle to free the social order from some of its earthly constraints in a new environment. But it is a struggle that can only ever partially succeed for precisely the sorts of reasons that Marx set out in his open letter to the Icarians (see above, Chapter 2).

The subterranean comparisons in the novel to the historical-geographical experience of post-enlightenment capitalism is daring (the new Martian social order arises in a way that loosely parallels the *Manifesto*'s account of the rise to power of the bourgeoisie). This could make Robinson's trilogy an easy target for highly critical post-colonial deconstruction. But I prefer to assess it in a more positive way. It holds out the tantalizing prospect of an inner connexion between actual historical-geographical transformations (understood with all the power that a properly constituted historical-geographical materialism can command) and the utopian design of an alternative spatiotemporal dynamics to that which we now experience.

3 The historical-geography of spatiotemporal utopianism

On one point, at least, Marx and Unger make common cause. They both insist that the future must be constructed, not in some fantastic utopian mold, but through tangible transformations of the raw materials given to us in our present state. Those raw materials were constructed and assembled through a spatiotemporal dynamics that was, however, inspired by a distinctive set of conflicting visions (including the overwhelming vision of capitalist and merchant entrepreneurs). A study of the historical geography of capitalism therefore provides clues as to how a spatiotemporal utopian project can be grounded in both the present and the past. This is obviously a task well beyond the confines of this book and it takes the notion of utopian thinking into a different dimension. But let me sketch in the sort of argument I have in mind.

Consider, for example, how free-market utopianism (the process) was put into place globally (geopolitically as spatial form) after World War II. In this the United States had an all-powerful, but specifically situated and particularistic, role. It was the epicenter from which a geopolitical strategy of global domination via freedom of the market was mobilized. It saw the dismantling of empires and decolonization, the shaping of the proper mediating international institutions (managed to ensure its own particular interests became the universal norm), and the opening of international trade as absolutely essential to the creation of a new world order. It saw itself in a life-and-death struggle with communism. Its self-image was as a beacon of freedom, individual rights, and democracy in a troubled world, as a model society to which everyone aspired, as a 'shining city on a hill' doing battle, as Ronald Reagan framed it, with an 'Evil Empire' of communism, as well as with the dark forces of ignorance, superstition, and irrationality. A secularized and more open spatiotemporality had to be imposed upon the world at a variety of scales (urban and regional as well as international), within which capital investments could more easily flow, along with movements of information, people, commodities, cultural forms, and the like. Nation and local states had to be built up as facilitators for freely functioning capital markets (executive committees, as in Baltimore City, for capital accumulation). This meant an attempt (often abortive) to impose (with a good deal of militarism and violence on the international stage) a particular conception of 'political democracy' (voting between political parties on a four- or five-year cycle) as a universal principle (as if there are no other possible ways of being free and democratic). The world's spaces were forced open through often violent struggles and then re-shaped by the power of US policies (including those of satellite states, comprador classes, and international institutions). Many of those engaged upon this project within the United States (of both left and right political persuasions, including many non-government organizations) deeply believed they were involved in a struggle to create a happier, more open, and freer world. They pursued with utopian conviction policies of development, aid, secular and military assistance, and education as means towards a humanistically powered enlightenment around the globe.

While this caricatures somewhat, it captures something important about the spatiotemporal utopianism of US internationalism over the last half century (a view subsequently given a Hegelian gloss in Fukuyama's 'end of history' thesis). It illustrates the *possibility* of a spatiotemporal utopian mode and gives a sense of what might be involved. By calling it a spatiotemporal utopianism we can better understand how it worked, why and how it went wrong, and how its internal contradictions might form one potential seedbed for some alternative.

Of course, events in the world did not dance solely to this vision. But scrutiny of the internal contradictions in this project is helpful. If the seeds of revolutionary transformation must be found in the present and if no society can launch upon a task of radical reorganization for which it is not at least partially prepared, then those internal contradictions provide raw materials for growing an alternative. Let me list some of the main contradictions:

1. The secular project of increasing material well-being throughout the whole world by means of extended capital accumulation failed to deliver on its promises. It could not satisfy human wants, needs, and desires nor liberate time and space for emotional and intellectual development. It promised unlimited consumerism (sometimes even as a collective good in terms of public welfare) as a path for the pursuit of happiness but delivered at best lopsided and at worst fraudulent benefits. It produced substantial wealth and empowerment for the few and disillusionment, repression, misery, and degradation for the rest. Its utopian claims respecting equality and well-being therefore came increasingly into contradiction with realities as one 'development decade' passed into another and as the uneven qualities of capitalism's geography became more and more apparent at a variety of scales (urban, regional, and international).

2. The promise of individual rights, freedoms, and liberties (the liberal illusion) embedded in the (often-exported) institutions of liberal democracy produced plenty of egotistical calculation (as the *Manifesto* has it) but produced a freedom to dominate and exploit others who were kept free of political influence and power by a politics of unequal rewards if not downright marginalization. It also failed to acknowledge the collective bases (cultural as well as political) necessary for securing liberties and freedoms, producing either a society of private wealth and public squalor (to use Galbraith's telling formulation) or authoritarian (and in some cases thoroughly corrupt) political structures founded on repressive tolerance. In the United States one is free to spend and vote how one likes but it is impossible to secure elementary freedoms of the city (e.g. walking the streets at any time of day or night) whereas in Singapore one can walk the city but not oppose the government or, for that matter, even buy chewing gum.

3. The overall 'success' of this utopian project was predicated upon a preparedness to exercise authority and where necessary to resort to

means of violence and repression as a necessary path to a more general enlightenment (in this it could not avoid the problems of the classical utopian forms). The trauma of the Vietnam War and subsequent revelations about covert operations around the world tarnished the utopianism of the project and made it appear more and more as an exercise in the power politics of US Manifest Destiny as seen by an elite few in the United States rather than as a mass movement for global enlightenment. The project could never free itself from the political conditions and often self-serving policies that characterized its command center in Washington. While the United States could reserve the right to be judge and jury of international morality it would never submit itself to being judged by international institutions like the United Nations and the World Court nor would it sign any pact (such as those against genocide or those concerning 'crimes against humanity') that would make it liable to such international judgements.

4. The spatial libertarianism of market forces undermined static territorial structures and powers (even, to some degree, those located in the United States) and were ruthlessly transformative with respect to 'traditional' cultural forms. The countereffect has been a return to territoriality and national identity as a basis for politics and a penchant for reactionary exclusionism that potentially threatens the free market agenda. After the collapse of communism in particular, many more turned to religion and/or nation as the sole alternative identity. To many (including forces of resistance within the United States, as well as in Iran, India, and Guatemala, to cite just a few) it now seems as if there is a simple choice between the secular spatiotemporalities of the free market or the mythological timespace of religion and nationhood.

5. Market externalities (costs not captured by the price mechanism) generated a wide range of social, economic, and political difficulties. Chief among these were problems of indiscriminate resource use, habitat destruction, and a whole series of environmental difficulties that required urgent attention. The concept of 'sustainability,' evolved in part to confront such difficulties, though easily coopted, points to spatiotemporal horizons different from those of capital accumulation.

When we bundle these contradictions together the picture that emerges is of a globalization process, centered on the United States, in serious disarray. There exists a broad swathe of disaffection from the undoubted achievements of the spatiotemporal utopianism led by the United States after World War II. Alternative visions need to uncover how to deliver on the promises of considerable improvement in material well-being and democratic forms, without relying upon egotistical calculation, raw consumerism, and capital accumulation, how to develop the collective mechanisms and cultural forms requisite for self-realization outside of market forces and money power, and how to bring the social order into a better working relation with environmental and ecological conditions.

4 Utopianism now?

The broad rejection of utopianism over the past two decades or so should be understood as a collapse of *specific* utopian forms, both East and West. Communism has been broadly discredited as a utopian project and now neoliberalism is increasingly seen as a utopian project that cannot succeed. Insofar as the geopolitical strategy of the United States can be understood as a form of spatiotemporal utopianism, it, too, is less and less convincing. So, should we just let the whole idea of utopianism of any sort die an unmourned death? Or should we try to rekindle and reignite utopian passions once more as a means to galvanize socio-ecological change?

Marx opposed utopianism as he knew it. He savaged the utopias of spatial form and thoroughly deconstructed Adam Smith's utopianism of social process. Yet Marx passionately believed in the emancipatory potential of class struggle as *the* privileged path towards a happier life. And both he and Engels argued in the *Communist Manifesto* that there are historical moments when oppositional forces are in such an undeveloped state that 'fantastic pictures of future society' come to represent 'the first instinctive yearnings' for 'a general reconstruction of society.' The literature produced by the socialist utopians of the early nineteenth century contains a powerful and important critical element. In attacking 'every principle of existing society,' they provided 'the most valuable materials for the enlightenment of the working class.' Furthermore, 'the practical measures proposed' were helpful as landmarks in the struggle to abolish class distinctions. The ever-present danger, they argued, is that we will come to believe 'in the miraculous effects' (1952 edition, 91) of some utopian science.

There is a time and place in the ceaseless human endeavor to change the world, when alternative visions, no matter how fantastic, provide the grist for shaping powerful political forces for change. I believe we are precisely at such a moment. Utopian dreams in any case never entirely fade away. They are omnipresent as the hidden signifiers of our desires. Extracting them from the dark recesses of our minds and turning them into a political force for change may court the danger of the ultimate frustration of those desires. But better that, surely, than giving in to the degenerate utopianism of neoliberalism (and all those interests that give possibility such a bad press) and living in craven and supine fear of expressing and pursuing alternative desires at all.

A critical examination of utopianism reveals some important variants and distinctive difficulties within the genre. The stark contrast between spatial form and social process utopianism, for example, reveals some peculiar habits of mind in the handling of space and time in visionary social thought. But even spatial form utopianism does a disservice to spatiality, for it typically treats space as a container for social action and confines utopianism most typically to the scale of the city (it is no accident, therefore, that the most explicit connexion between social action and utopian thinking is at the urban scale). How spatial form utopianism would look under conditions of the dynamic production of space and in relation to a theory of uneven geographical developments (cf. Chapter 5) remains unexplored.

How, then, can a stronger utopianism be constructed that integrates social process and spatial form? Is it possible to formulate a more dia-lectical form of utopianism, construct, even, a utopian dialectics?

For this to happen requires a dialectics that can operate in relation to both space and time (something impossible within the Hegelian tradition). It also has to face up to the materialist problems of authority and closure. Closure (the making of something) of any sort contains its own authority because to materialize any one design, no matter how playfully construed, is to foreclose, in some cases temporarily but in other instances relatively permanently, on the possibility of materializing others. We cannot evade such choices. The dialectic is 'either/or' not 'both/and.' What the materialized utopianism of spatial form so clearly confronts is the proble-matics of closure and it is this which the utopianism of the social process so dangerously evades. Conversely we find that fragmentation and dispersal cannot work, and that the bitter struggle of the 'either-or' perpetually interferes with the gentler and more harmonious dialectic of 'both-and' when it comes to socio-ecological choices. We also find that the shadowy forms of spatiotemporal utopianism are not too hard to exhume from a study of our own historical geography as impelled by the geopolitics of capitalism. The task is then to define an alternative, not in terms of some static spatial form or even of some perfected emancipatory process. The task is to pull together a spatiotemporal utopianism – a dialectical utopianism – that is rooted in our present possibilities at the same time as it points towards different trajectories for human uneven geographical developments. It is to that task that I now turn.

PART 4

CONVERSATIONS ON THE PLURALITY OF ALTERNATIVES

CHAPTER 10

On architects, bees, and 'species being'

When, sometime in the early eighteenth century, Bernard Le Bovier Fontenelle wrote his *Conversations on the Plurality of Worlds*, he devised a novel way in which to try to persuade a sceptical audience of the possible truth of the Newtonian world view. The conversations, with an elegant and discerning lady, took place in the course of evening walks in a garden. In such a setting, it seemed possible to contemplate alternative possibilities away from the hurly-burly of daily life and gain thereby a different perspective on the world.

It is hard, in contemporary circumstances, to think of ways in which to conduct a similar conversation. Yet this is what we must do if we are to uncover possible alternatives to the social world we currently inhabit. In the absence of any obvious blueprint for social change (a blueprint which would in any case likely be dismissed as visionary nonsense), and in the absence (regrettable as it may be) of any major social movement or vigorous class alliance pressing forward immediate theses and plans for social change, the best that I can offer is a series of talking points around which conversations about alternatives and possibilities might coalesce.

This means, if the arguments of Part 3 are correct, coming to terms with something called 'dialectical utopianism.' Dialectics here denotes something different from that usually understood from studies of Hegel or even of Marx. It presumes, for example, a dialectics able to address spatio-temporal dynamics openly and directly and able also to represent the multiple intersecting material processes that so tightly imprison us in the fine-spun web of contemporary socio-ecological life. It then entails a willingness, if only in the world of thought, to transcend or overturn the socio-ecological forms imposed by uncontrolled capital accumulation, class privileges, and gross inequalities of political-economic power. In this way, a space for thought experiments about alternative possible worlds can be constructed. While there is always a danger that this might degenerate into the production of unrealizable dreams, getting the historical and geographical materialism right should help convert those dreams into prospects that really do matter.

1 On architects and bees

I begin with the figure of the architect. I do so in part because that figure (and it is the *figure* rather than the professional person of whom I speak) has a certain centrality and positionality in all discussions of the processes of constructing and organizing spaces. The architect has been most deeply enmeshed throughout history in the production and pursuit of utopian ideals (particularly though not solely those of spatial form). The architect shapes spaces so as to give them social utility as well as human and aesthetic/symbolic meanings. The architect shapes and preserves long-term social memories and strives to give material form to the longings and desires of individuals and collectivities. The architect struggles to open spaces for new possibilities, for future forms of social life. For all of these reasons, as Karatani (1995, XXXV) points out, the 'will to architecture' understood as 'the will to create' is 'the foundation of Western thought.' Plato held to that view and Leibniz even went so far as to say: 'God as architect fully satisfies God as lawgiver.'

But the other reason I insist on the *figure* of the architect is because there is a sense in which we can all equally well see ourselves as architects of a sort. To construe ourselves as 'architects of our own fates and fortunes' is to adopt the figure of the architect as a metaphor for our own agency as we go about our daily practices and through them effectively preserve, construct, and re-construct our life-world. This reconnects directly to Marx. For it is hard to find a better statement of the foundational principles of the dynamics and dialectics of socio-ecological change than those laid out in the first volume of *Capital* (1967 edition, 177–8):

> Labour is, in the first place, a process in which both man and Nature participate, and in which man of his own accord starts, regulates, and controls the material re-actions between himself and Nature ... By thus acting on the external world and changing it, he at the same time changes his own nature. He develops his slumbering powers and compels them to act in obedience to his sway ... We presuppose labour in a form that stamps it as exclusively human. A spider conducts operations that resemble those of a weaver and a bee puts to shame many an architect in the construction of her cells. But what distinguishes the worst architect from the best of bees is this, that the architect raises his structure in imagination before he erects it in reality. At the end of every labour process we get a result that existed in the imagination of the labourer at its commencement. He not only effects a change of form in the material on which he works, but he also realizes a purpose ...

The parallel with Park's conception of urbanization (see Chapter 8) is uncanny. More important for the present argument is that Marx's analogy can easily be reversed: while the activities of the architect help us understand the labor process in general, everyone who engages in any kind of labor process whatsoever is like the architect rather than like the bee.

Marx's evocation of bees has, however, a double reckoning. Not only does it relate directly to the sophistication of their architectural practices (so fascinating to naturalists), but Marx is undoubtedly also referring to Mandeville's famous tract of 1714 on *The Fable of the Bees* (with its subtitle of 'private vices and public virtues'). Mandeville not only commented therein on how public prosperity and virtue necessarily rested on private vanity, envy, vice, and waste (a problem that Adam Smith's utopianism of the market was later designed to redress) but took the even more invidious though 'honest' position (Marx 1976 edition, 764–5) that society in general could prosper only if the workers remained poor, ignorant, and deprived of any and all knowledge that might multiply their desires. Marx's concept of human labor in general is obviously meant to contrast with this idea of the ignoble and degraded status of a 'worker bee' under capitalism. The latter obviously has little or no chance to awaken those 'slumbering powers' latent within us to change the world and change ourselves.

We now know a lot more about bees. They are, for example, very communicative creatures. The dance choreography they perform in the hive provides precise information as to where food sources can be found. The intricacy and complexity of the communication system (and the accuracy and precision incorporated in it) demonstrates a truly amazing capacity for bees to encode and communicate information in an abstract, symbolic way that would put to shame many a communications or GIS specialist let alone any architect (Von Frisch, 1965, took forty years to map the dances). The code to the dance patterns was broken, almost by accident, by a mathematician who happened to be the daughter of a bee researcher. She recognized the patterns when projecting the properties of a six-dimensional flag manifold – a rare and obscure kind of mathematics – onto a two-dimensional space (Frank, 1997). The entire repertory of bee dances with all of its innumerable parts and variations falls within a mathematical schema unknown to any architect. The only other known physical process to which such a mathematics applies concerns the quarks of quantum theory. This raises the speculative possibility that 'the bees are somehow sensitive to what's going on in the quantum world of quarks, that quantum mechanics is as important to their perception of the world as sight, sound, and smell' (86). If this turns out to be true, then not only do bees 'know' (with a tiny brain) a kind of mathematics known to only a handful of people, but they also may be able to do what no human appears ever able to do – operate in quantum fields without disturbing them. So, even as we enter the age of quantum computing with all of its untold power, we still cannot do what bees seem able to do.

The more we know about bees, the more the comparison with even the best of human labor (let alone the worst of architects) appears less and less complimentary to our supposedly superior powers. This seriously dents any idea that humans are somehow at the 'summit' of living things in all or even most respects. But it also sharpens interest in the question of what our 'exclusive' species capacities and 'slumbering powers' might be.

2 Human capacities and powers

Many species, like bees, possess 'basic senses entirely outside the human repertory.' From this, Wilson (1998, 47–8) formulates 'an informal rule of biological evolution important to the understanding of the human condition: If an organic sensor can be imagined that picks up any signal from the environment, there exists a species somewhere that possesses it.' It is not surprising, therefore, that the unaided human senses we possess 'seem remarkably deficient relative to the bountiful powers of life expressed in such diversity.' Wilson's characteristically reductive answer to why this is so runs as follows:

> Biological capacity evolves until it maximizes the fitness of organisms for the niches they fill, and not a squiggle more. Every species, every kind of butterfly, bat, fish and primate, including *Homo sapiens*, occupies a distinctive niche. It follows that each species lives in its own sensory world.

When, therefore, we appeal as we did in Chapter 6 to the idea of 'the body as the measure of all things' we immediately encounter the limitations of our own sensory world. But human beings have acquired means to 'listen, see and hear' far beyond such limitations. Our capacities as 'cyborgs and scientists' cannot be ignored. This poses a fundamental problem for both Marx and Wilson (unlikely allies both) in their search for some sort of unity of knowledge. Wilson's version of it is this:

> Natural selection [cannot] anticipate future needs ... If the principle is universally true, how did natural selection prepare the mind for civilization before civilization existed? That is the great mystery of human evolution: how to account for calculus and Mozart.   (48)

This is a familiar problem in Marx. In innumerable passages, from the *Communist Manifesto* on, he appears to contradict the conception of the labor process laid out in *Capital* and insist that our ideas, conceptions, views (in one word, our 'consciousness') change with every change in material conditions of existence and that the material form of a mode of production gives rise to institutional, legal, and political structures which imprison our thoughts and possibilities in particular ways. In perhaps the most famous rendition of this, Marx argues 'it is not the consciousness of men that determines their being but, on the contrary, their social being that determines their consciousness' (Marx and Engels, 1972 edition, 4). How, then, can the human imagination, made so much of in *Capital,* range freely enough outside of the existing material and institutional conditions (e.g. those set by capitalism) to even conceptualize what the socialist alternative might look like? In exactly the same way that Wilson has a problem accounting for the explosion of cultural and scientific forms in recent history, so Marx's historical materialism has a problem in preparing our imaginations (let alone our political practices) for the creation of a socialist (or for that matter any other) alternative.

While this may explain how we can be 'such puppets of the institutional and imaginative worlds we inhabit' (to repeat Unger's trenchant phrase) it also presents a difficult paradox. The historical-geographical experience of revolutionary movements in power (and of materialized utopianism of any sort) indicates the deep seriousness of the problem of unpreparedness for radical change. Many revolutionary movements did not or could not free themselves from ways of thinking embedded in the material circumstances of their past. The dilemma is as pertinent and real in political practices as it is salient theoretically. Unger's thought perpetually gravitates back to this central problem. It is a fundamental dilemma that any grounded form of dialectical utopianism must confront.

Marx (1970 edition, 20–1) does, however, soften the theoretical paradox somewhat:

> At a certain stage of their development, the material forces of production come in conflict with the existing relations of production . . . From forms of development of the forces of production these relations turn into their fetters. Then begins an epoch of social revolution. With the change of the economic foundation the entire immense superstructure is more or less rapidly transformed. In considering such transformations a distinction should always be made between the material transformation of the economic conditions of production which can be determined with the precision of natural science, and the legal, political, religious, aesthetic or philosophic – in short ideological forms in which men become conscious of this conflict and fight it out.

The latter (the ideological forms) do not here appear to be as strictly determined by material conditions as initially proposed (in part because of their inherent fuzziness) while the very existence of contradictions (particularly between the forces and relations of production) holds out the possibility for creative maneuver and open decision-making.

Nevertheless we often seem to oscillate in our understandings of ourselves and in our ways of thinking between an unreal fantasy of infinite choice (Unger's 'alternatives that scarcely seem to matter') and a cold reality of no alternative to the business as usual dictated by our material and intellectual circumstances.

This is why the figure of the architect is so instructive. Consider it further. It takes a huge exercise of the imagination to design an office tower, a residence, a factory, a leisure park, a city, or whatever. The architect has to imagine spaces, orderings, materials, aesthetic effects, relations to environments, and deal at the same time with the more mundane issues of plumbing, heating, electric cables, lighting, and the like. The architect is not a totally free agent in this. Not only do the quantities and qualities of available materials and the nature of sites constrain choices but educational traditions and learned practices channel thought. Regulations, costs, rates of return, clients' preferences, all have to be considered to the point where it often seems that the developers, the financiers, the accountants, the builders, and the state apparatus have more to say about the final shape of things than the architect. The process of 'doing architecture' entails all these complications. 'Doing architecture' is an embedded, spatiotemporal practice. But there is, nevertheless, always a moment when the free play of the imagination – the will to create – must enter.

The inner connection at work within Marx's oppositional statements then becomes more understandable. All capitalist ventures, including those of the architect, are speculative. This is what it means to throw money into circulation as capital and hope to realize a profit. All capitalist ventures must exist in the imagination before they are realized in the market (hence the acknowledged power of human expectations in economic action). The incredible power of capitalism as a social system lies in its capacity to mobilize the multiple imaginaries of entrepreneurs, financiers, developers, artists, architects, and even state planners and bureaucrats (and a whole host of others including, of course, the ordinary laborer) to engage in material activities that keep the system reproducing itself, albeit on an expanding scale. The discipline – such as it is – imposed by the system comes through the acid test of profitability. It is only then that the imaginary realizes itself in ways that gain positive reinforcement. But there are as many ways to make a profit as to skin a proverbial cat. So while the singular goal of profit may guide capitalist activity, there is no single path to reach that goal. Indeed, the whole history of the capitalist imaginary has been to find all sorts of innovative and often quirky ways to realize that singular objective. Giving free rein to the imagination is fundamental to the perpetuation of capitalism and it is within this space that an alternative socialist imaginary can grow (though not now in a manner that is disembedded from capitalism and its dominant ways of thinking and doing).

What we then recognize is a simple material fact about the way our world, the world of capitalist culture, economy, politics, and consciousness, works. It is full of an incredible variety of imagined schemes (political, economic, institutional), many of which get constructed. Some schemes fail. Others are wildly successful. Some work for a time and then fall apart. It is the cold logic of the market place (often lubricated by a hefty dose of political favoritism and conniving collusions) that fixes the success or failure of the outcome. But it is engagement with future possibilities that starts the whole affair. Zola captured this idea beautifully in his exposé of the power of money to transform the world through speculation. Says Saccard, Zola's anti-hero in his novel *Money*:

> [Y]ou will behold a complete resurrection over all those depopulated plains, those deserted passes, which our railways will traverse – yes! fields will be cleared, roads and canals built, new cities will spring from the soil, life will return as it returns to a sick body, when we stimulate the system by injecting new blood into exhausted veins. Yes! money will work these miracles . . . You must understand that speculation, gambling, is the central mechanism, the heart itself, of a vast affair like ours. Yes, it attracts blood, takes it from every source in little streamlets, collects it, sends it back in rivers in all directions, and establishes an enormous circulation of money, which is the very life of great enterprises . . . Speculation – why, it is the one inducement that we have to live; it is the eternal desire that compels us to live and struggle. Without speculation, my dear friend, there would be no business of any kind . . . It is the same as in love. In love as in speculation there is much filth; in love also, people think only of their own gratification; yet without love there would be no life, and the world would come to an end.
>
> (Zola, 1891, 140)

Saccard's vision, his love of life, seduces all around him. Even his cautious and demure companion – Mme Caroline – is struck by how the present state of the land in the Levant fails to match up with human desires and potentialities:

> And her love of life, her ever-buoyant hopefulness, filled her with enthusiasm at the idea of the all-powerful magic wand with which science and speculation could strike this old sleeping soil and suddenly reawaken it . . . And it was just this that she saw rising again – the forward, irresistible march, the social impulse towards the greatest possible sum of happiness, the need of action, of going ahead, without knowing exactly whither . . . and

amid it all there was the globe turned upside down by the ant-swarm
rebuilding its abode, its work never ending, fresh sources of enjoyment ever
being discovered, man's power increasing ten-fold, the earth belonging to
him more and more every day. Money, aiding science, yielded progress. (75)

While the outcome suggests there is no alternative, the starting point holds
that there are at least a million and one alternatives as we seek to probe
future possibilities with all the passion and imagination at our command.
The dialectic of the imaginary and its material realization (mediated in
most instances through production) locates the two sides of how capitalism
replicates and changes itself, how it can be such a revolutionary mode of
production. Capitalism is nothing more than a gigantic speculative system,
powered, as Keynes for one clearly recognized, by some mix of 'expectations' (respectable) and 'speculative behaviour' (disreputable). If such
fictitious and imaginary elements surround us at every turn, then the
possibility also exists of 'growing' imaginary alternatives within its midst.

Marx did not object to the utopian socialists because they believed that
ideas could be a material force in historical change but because of the way
they derived and promoted their ideas. Plucked from some rarefied ether
of the imagination, such ideas were doomed to failure. Extracted from the
womb of bourgeois society, or, as Zola might put it, from the 'fertile
dungheap' of its contradictions, ideas could provide the basis for a transformative politics. The working class 'have no ideal to realise,' Marx
(Marx and Engels 1972 edition, 558) wrote in his commentary on *The
Civil War in France*, 'but to set free the elements of the new society with
which old collapsing bourgeois society itself is pregnant.' It is the task of
dialectical and intellectual enquiry to uncover real possibilities and alternatives. This is where a dialectical utopianism must begin.

3 The conception of 'our species being'

To speak of our capacities to transform the world through labor and
thereby to transform ourselves, and to speak also of how we might deploy
our albeit constrained imaginations in such a project, is to presuppose
some way of understanding ourselves as a species, our specific capacities
and powers (including the 'slumbering powers' of which Marx speaks) in
relation to the world we inhabit. The dialectical and metabolic relation we
have to nature and through that back to a distinctively human nature (with
its special qualities and meanings) must therefore lie at the basis of what
we, as architects of our futures and our fates, can and want to accomplish.

Serious problems have arisen in social theory as well as in the quest for
alternatives whenever a biological basis – such as that invoked in a concept

like 'species being' – has been invoked (familiar examples include the way the arguments of social Darwinism were incorporated into Nazism, organicist theories of the state, the dismal history of the eugenics movement particularly as applied to racial categories, and the profound social antagonisms generated in the debate over sociobiology during the 1970s). Much of the writing in this genre has indeed been reactionary, conservative, and fatalistic with a strong dash of biological determinism (these days usually genetic) thrown in. The general response on the social science side and throughout much of the left in recent times has been to retreat from any examination of the biological/physical basis of human behavior. Within Marxism, for example, the trend has been to treat human nature as relative to the mode of production (or to material life in general) and to deny any universal qualities to our species being.

This is not, as Geras (1983) expertly argues, an adequate response (nor is it at all consistent with Marx's formulations). Unless we confront the idea, however dangerous, of our human nature and species being and get some understanding of them, we cannot know what it is we might be alienated from or what emancipation might mean. Nor can we determine which of our 'slumbering powers' must be awakened to achieve emancipatory goals. A working definition of human nature, however tentative and insecure, is a necessary step in the search for real as opposed to fantastic alternatives. A conversation about our 'species being' is desperately called for.

I propose a basic conception that goes roughly like this. We are a species on earth like any other, endowed, like any other, with specific capacities and powers that are put to use to modify environments in ways that are conducive to our own sustenance and reproduction. In this we are no different from all other species (like termites, bees, and beavers) that modify their environments while adapting further to the environments they themselves help construct.

This conception defines 'the nature imposed condition of our existence.' We are sensory beings in a metabolic relation to the world around us. We modify that world and in so doing change ourselves through our activities and labors. Like all other species, we have some species-specific capacities and powers, arguably the most important of which are our ability to alter and adapt our forms of social organization (to create, for example, divisions of labor, class structures, and institutions), to build a long historical memory through language, to accumulate knowledge and understandings that are collectively available to us as guides to future action, to reflect on what we have done and do in ways that permit learning from experience (not only our own but also that of others), and, by virtue of our particular dexterities, to build all kinds of adjuncts (e.g.

tools, technologies, organizational forms, and communications systems) to enhance our capacities to see, hear, and feel way beyond the physiological limitations given by our own bodily constitution. The effect is to make the speed and scale of adaptation to and transformation of our species being and of our species environment highly sensitive to the pace and direction of cultural, technological, economic, social, and political changes. It is this, of course, that makes so much (though not all) of what we think and do subservient to the inherent dynamics of a dominant mode of production. The argument for seeing human nature in relative terms, as something in the course of construction, is not without weight and foundation. But it also points to a connection between the concept of 'species being' and 'species potential.'

We can never get away from the universal character of our existence as sensory and natural beings, the product of a biological and historical-geographical evolutionary process that has left its mark upon our species both in terms of genetic endowments and rapidly accumulating cultural acquisitions. The sociobiologists are right to insist upon the significance of our genetic heritage. No conception of human nature can ignore what modern genetics and microbiology are revealing about human constraints, capacities, and powers. The collapse of the Cartesian dualism of mind versus matter through contemporary studies of the mind/brain problem is likewise leading the way to a radical reformulation of the relation between thought and action in human behaviors.

Sociobiology provides no adequate explanation of cultural and social evolution, especially that of recent times. While it is plausible to argue for some kind of co-evolution between biological characteristics and cultural forms over the long term, the explosion of cultural/technical/linguistic understandings and practices particularly over the last 300 years has provided no time for biological adaptation. It has, furthermore, no possible causative or reductive explanation in terms of physical or biological processes alone. The latter may form the necessary foundations for socio-ecological change but they cannot provide sufficient explanations for the rise of civilizations (let alone for calculus and Mozart). In effect, the situation we have to confront is one in which genetic endowments have been put to use in entirely new cultural ways. But what are these endowments that provide the raw materials out of which we are fashioning our historical geography?

We are, at root, curious and transformative beings endowed with vivid imaginations and a certain repertoire of possibilities that we have learned to put together in different ways at different places and times. We are political and semiotic animals with respect to each other, and politics is grounded in communicative abilities that are themselves evolving rapidly.

Among our more endearing habits, furthermore, is the ability to be sophisticated rule makers and compulsive rule breakers. Indeed, a case can be made (and here I parallel the general thrust of Unger's work) that emancipation is best defined by a condition in which we can be both rule makers and rule breakers with reasonable impunity (for this reason Unger considers what he calls 'immunity rights' to be a fundamental feature of any society that aspires to emancipatory forms of development). But the rule making has to acknowledge a bundle of constraints and possibilities derived from our distinctive and achieved metabolic condition.

The basic repertoire derived from evolutionary experience provides strategic options for human action. The repertoire includes:

1. competition and the struggle for existence (the production of *hierarchy and homogeneity* through natural or, in human history, economic, political, and cultural selection);

2. adaptation and diversification into environmental niches (the production of *diversity* through proliferation and innovation in economic, political, or cultural terms);

3. collaboration, cooperation, and mutual aid (the production of *social organization*, institutional arrangements, and consensual political-discursive forms, all of which rest upon capacities to communicate and translate);

4. environmental transformations (the transformation and modification of 'nature' into, in our case, a *humanised nature* broadly in accord – though with frequent unintended consequences – with human requirements);

5. spatial orderings (mobilities and migrations coupled with the *production of spaces* for distinctive purposes such as escape, defense, organizational consolidation, transport, and communication, and the organization of the spatially articulated material support system for the life of individuals, collectivities, and the species); and

6. temporal orderings (the setting up of biological, social, and cultural 'clocks' that contribute to survival coupled with the use of various time orderings for biological and social purposes – in human societies time orderings vary from the almost instantaneous transmission of computerized orders to the long-term contracts that evolve by culture into moral precepts, tradition, and law).

These six elements form a basic repertoire of capacities and powers handed down to us out of our evolutionary experience. When faced with a difficulty we have choices. Put crudely, we can stand and fight, defuse the difficulty by diversifying into something non-competitive, cooperate, change the environmental conditions that give rise to the problem, move out of the way, or put ourselves on a different time horizon (e.g. delay and defer into the future).

While all organisms may possess some or even all of the repertoire in some degree, there is no question that human beings have highlighted each element in particular ways (e.g. the long-term temporal relations of culturally transmitted traditions) and achieved rich and flexible ways of combining the different elements into complex social systems. Each mode of production can be construed as a special combinatorial mix of elements drawn from this basic repertoire.

But it is vital to interpret the categories relationally rather than as mutually exclusive (see Harvey, 1996, for a fuller statement). I think sociobiologists are correct, for example, to argue that cooperation ('reciprocal altruism' is their preferred term) is in some sense an adaptive form of competition (organisms that help each other survive better). The difficulty arises when they make the competitive moment the foundation of everything else (a convenient way to make capitalist competition appear as *the* fundamental law of nature). From a relational standpoint, competition can just as easily be seen as a form of cooperation. The production of territoriality is an interesting case in point. By defining territories competitively, organisms cooperatively organize the partition of resources to save on ruinous and destructive competition. Properly organized, territoriality is as much about collaboration in human affairs as it is about competition and exclusion.

The character of a social formation is defined by exactly how the elements in the overall repertoire get elaborated upon and combined through the exigencies of class power. Capitalism, for example, is often construed as being basically about competition. Survival of the fittest (measured in terms of profitability) is the Darwinian mechanism that creates order out of the chaos of speculative and competitive economic activity. But capitalism is also highly adaptive, constantly searching out innovative strategies, new market niches, and new product lines precisely to avoid competition in already established fields. Furthermore, capitalism could not survive without a lot of cooperation, collaboration, and mutual aid. I speak here not only of the ways in which supposed competitors so frequently collude (clandestinely or overtly) or of the extensive arenas of social organization (such as air traffic control) that rely upon tight submission of any competitive instincts to organized social control, but also of the extensive regulatory mechanisms embedded primarily in state power and the law to ensure that markets function as a consensual and collaborative framework for competition. The transformation or 'production' of nature through collaborative efforts (in, say, the fields of plant and animal breeding, and now genetic engineering, the construction of physical infrastructures, the building of cities, and the like) generates rapidly evolving environments (both social and physical) within which different forms of competitive, adaptive, or collaborative behavior can arise. Uneven geographical developments shape entirely new market niches, for example. And, as I have often emphasized, capitalism has found remarkable ways to produce new spatial configurations, to measure and coordinate turnover times, and thereby construct entirely different spatiotemporalities to frame its own activities.

It is not, therefore, competition alone that defines capitalism, but the particular *mode* of competition as embedded in all the other evolutionary processes. Institutions, rules, and regulations struggle to ensure that only one sort of competition – that within relatively freely functioning markets respecting property rights and freedom of contract – will prevail. The normal causal ordering given in sociobiology can easily be reversed; it is only through the collaborative and cooperative structures of society (however coerced) that competition and the struggle for existence can be orchestrated to do its work (and it is notable how often capitalists complain of 'ruinous competition' and call immediately for government regulation to cure the problem). Without the extensive networks of collaboration and cooperation already in existence, most of us would be dead. Competition is always regulated and conditioned by cooperation, adaptation, environmental transformations, and through the production of space and time.

This illuminates how an alternative to capitalism might begin to be construed. The traditional way of thinking about socialism/communism, for example, is in terms of a total shift from, say, competition to cooperation, collaboration, and mutual aid. This is far too simplistic and restrictive. If capitalism cannot survive without deploying all of the repertoire in some way, then the task for socialism must be to find a different combination of *all* the elements from within the basic repertoire. This cannot be done by presuming that only one of the elements matters and that the others can be suppressed. Competition, for example, can never be eliminated. But it can be organized differently and with different ends and goals in view. The balance between competition and cooperation can be altered. This has frequently occurred in capitalist history as phases of 'excessive competition' alternate with phases of strong state regulation. The recent move towards globalization is an example of how a shift in one key element in the repertoire – the production of space – can occur in the struggle to sustain the system.

The history of socialist theory is full of debates on the feasibility of this or that form of social organization. Recent arguments have focussed on whether 'market socialism' is more achievable or desirable than democratically controlled central planning. The lively debate brought together by

Ollman (1998) revolves around these issues, but it is curious that not one of the participants cares to set his arguments in the context of species capacities and powers. Had they done so, they might have seen a quite different set of possibilities.

Any mode of production is a contradictory and dynamic unity of different elements drawn from the basic repertoire I have outlined. There is plenty of contradiction, tension, and conflict within it, and these provide a set of embedded possibilities to construct alternatives. The transition from one mode of production to another entails transformations in all elements in the repertoire in relation to each other. 'No natural laws can be done away with,' Marx wrote in a letter to Kugelmann in 1868, adding, 'what can change, in historically different circumstances, is only the form in which these laws operate.' Or, just as persuasively, 'no social order can accomplish transformations for which it is not already internally prepared.'

All species (including human beings) can affect subsequent evolution through their behavior. All species (including humans) make active choices and by their behavior change the physical and social conditions with which their descendants have to cope. They also modify their behavior in response to changed conditions, and by moving expose them-selves to new conditions that open up different possibilities for evolu-tionary change. Organisms 'are not simply *objects* of the laws of nature, altering themselves to bend to the inevitable, but active *subjects* transform-ing nature according to its laws' (Lewontin, 1982, 162). Here the concept of 'species potential' returns to the fore because we are now more than ever architects of evolution by virtue of the scientific, technical, and cultural powers that we have acquired. We are not, nor can we ever be, master achitects so close to God as to be the ultimate lawgiver (as Leibniz evidently thought). But we have worked ourselves into a position in which the future of all evolution, including our own, is as much a function of conscious political and social choices as it is of random events to which we respond. So what kind of evolution do we, as savvy architects, imagine and plan? The answer depends on how we recombine the elements of the repertoire.

CHAPTER 11

Responsibilities towards nature and human nature

Consider just one element in the repertoire of our evolutionary capacities: the kind of 'nature' we are now in a position to produce. As active subjects in the evolutionary game we have accumulated massive powers to transform the world. The way we exercise those powers is fundamental to the definition of what we as a species will become. This is now an open and critical focus of discussion and debate, as much among the capitalists and their allies (many of whom are obsessed with the issue of long-term sustainability) as among those who seek alternatives. Do we have a distinctive 'species being' and what does that entail about our future relation to external nature?

The concept of 'species being' is, of course, species centered. It entails a resolutely anthropocentric stance. We cannot ever avoid (any more than can bees and beavers) asserting our own species identity, being expressive of who we are and what we can become, and putting our species capacities and powers to work in the world we inhabit. To construe the matter any other way is to fool ourselves (alienate ourselves) as to who and what we are.

What partially separates us human architects from bees, however, is that we are now obliged (by our own achievements) to work out in the imagination as well as through discursive debates our individual and collective responsibilities not only to ourselves and to each other but also to all those other 'others' that comprise what we usually refer to as 'external' nature ('external,' that is, to us). We have reached an evolutionary condition in which conscious choices can and need to be made not only about our own evolutionary paths but also about those of other species. Even genetic evolution, says Wilson (1998, 270), 'is about to become conscious and volitional, and usher in a new epoch in the history of life' for which, unfortunately (according to his schema), we are not genetically prepared. We have long been powerful evolutionary agents through everything from plant and animal breeding, massive habitat modification, and rapid population growth to the diffusion and mixing of species on a global scale. But we have been rapidly accumulating far vaster powers over the last two centuries.

This in no way means we are somehow 'outside of' metabolic or evolutionary constraints or invulnerable to natural forces. But we are in a position to consciously deploy the repertoire of evolved possibilities in radically different combinatorial ways. If the full volitional period of evolution is about to begin, then, says Wilson (1998, 277), in a statement that smacks far more of traditional humanism than biological reductionism, 'soon we must look deep within ourselves and decide what we wish to become.' That question is as deeply speculative about our species being and our species destiny as anything that any architect has ever faced before.

1 The discourses of nature

Grappling with responsibilities and ethical engagements towards all others entails the construction of discursive regimes, systems of knowledge, and ways of thinking that come together to define a different kind of imaginary and different modes of action from those which Zola, for example, depicted as so typical of the capitalist entrepreneur determined to master nature and construct a world in his own image. But on what basis do we seek to construct some alternative? There is no lack of passionate advocacy of this or that solution – the environmental and ecological movements are full of competing and cacophonous claims as to the possible future of the human species on planet earth.

Consider some of the major axes of difference. Ecocentric or biocentric views vie with naked anthropocentrism. Individualism clashes with collectivism (communitarianism). Culturally and historically-geographically embedded views (particularly those of indigenous peoples) sit uneasily alongside universal claims and principles (often advanced by scientists). Broadly materialist and economistic concerns over access to life chances (whether it be of species, individuals, social groups, or habitats) are frequently opposed to aesthetic, spiritual, and religious readings. Hubristic attitudes of promethean domination contrast with humility before the mighty and wondrous powers of nature. Innumerable villains (enlightenment reason, speciesism, modernity and modernization, scientific/technical rationality, materialism [in both the narrow and broader sense], technological change [progress], multinationals [particularly oil], the world bank, patriarchy, capitalism, the free market, private property, consumerism [usually of the supposedly mindless sort], state power, imperialism, state socialism, meddling and bumbling bureaucrats, military industrial complexes, human ignorance, indifference, arrogance, myopia and stupidity, and the like) all jostle (singly or in some particular combination) for the position of arch-enemy of ecological sanity. And the long-standing debate over ends versus means (authoritarian, democratic, managerial, personal) has plenty of echoes in environmental politics.

I know these are caricatures of some of the binary positions to be found within the environmental/ecological movement. They are, however, the sorts of oppositions that make for innumerable confusions, particularly when taken in combination. But there are added complications. Nobody knows, for example, exactly where an 'ecosystem' or a 'community' (the usual unit of analysis) begins and ends. Claims worked out and agreed upon at one geographical scale (the local, the bioregion, the nation) do not necessarily make sense when aggregated up to some other scale (e.g. the globe). What makes sense, furthermore, for one generation will not necessarily be helpful to another. And every political movement under the sun – from Nazis to free-market liberals, from feminists to social ecologists, from capitalists to socialists, from religious fundamentalists to atheistic scientists – necessarily considers it has the exclusive and correct line on environmental issues because to be seen as 'natural' is to assume the mantle of inevitability and probity.

Put all this together and we have a witches' brew of political arguments, concepts and difficulties that can conveniently be the basis of endless academic, intellectual, theoretical, and philosophical debate. There is enough grist here to engage participants at learned conferences until kingdom come. Which, in a way, is what makes the whole topic so intellectually interesting. To speak of consensus (or even sketch it as a goal) is plainly impossible in such a situation. Yet some common language, or at least an adequate way of translating between different languages (scientific, managerial and legal, popular, critical, etc.), is required if any kind of conversation about alternatives is to take place. Even in the midst of all this conflict and diversity, therefore, some sort of common grounding must be constructed. Without it, authoritarianism, discursive violence, and hegemonic practices become the basis for decisions and this, of course, is unlikely to create space for alternative possibilities.

2 Metaphors for survival

There are, however, some dominant simplifying metaphors to help guide deliberations. Such metaphors, while indispensable, have their pitfalls. Consider, for example, the overwhelmingly powerful role now played by ideas of environmental crisis, imminent ecological collapse, or even 'the end of nature' in oppositional environmental thinking. For some on the left, this rhetoric attracts because it has the convenient effect of displacing a long-standing belief in the ultimate crisis and collapse of capitalism from the field of class struggle to that of the environment. An alternative *must* be found immediately, the argument goes, if the world as we know it is not to end in environmental catastrophe. This is the crucible out of which entirely new alternative social forms must be forged in the near future.

Apocalyptic argument of this sort is not confined to extreme environmentalists. Many scientists sound similarly alarmist calls. In a declaration signed by more than two thousand of the world's most prestigious scientists we read:

> Human beings and the natural world are on a collision course. Human activities inflict harsh and often irreversible damage on the environment and on critical resources. If not checked, many of our current practices put at risk the future that we wish for human society and the plant and animal kingdoms, and may so alter the living world that it will be unable to sustain life in the manner that we know. Fundamental changes are urgent if we are to avoid the collision our present course will bring about.
>
> (Union of Concerned Scientists, 1996)

This language of 'humanity on a collision course with the natural world' is odd in many respects. Making it seem as if human beings are somehow outside of nature, it turns humanity into the metaphorical equivalent of some asteroid set to collide with the rest of nature, thereby avoiding the long history of evolutionary changes through which human beings have symbiotically transformed the world and themselves. The statement reeks of those 'abstract and ideological conceptions' of which Marx (1976 edition, 494) complained 'whenever (natural scientists) venture beyond the bounds of their own speciality.'

Such an alarmist rhetoric of crisis and imminent catastrophe is dangerous. To begin with, it presumes we know with utmost certitude and precision the flash point of some collision between 'human beings and the natural world.' But most scientists, even those who issue clarion calls for action, continuously hedge their bets as to where the really serious problems lie and how imminent they truly are. Wilson (1998, 285–7) thinks 'the wall toward which humanity is evidently rushing is a shortage not of minerals and energy, but of food and water.' The capacity of the earth to support its 'voracious human biomass' is, he suggests, becoming dicier and dicier. But others point to global warming and climate change, loss of habitats, and biodiversities (a theme also dear to Wilson's heart), degradation of a wide array of biosystems (from tropical rain forests to oceans), and the problems of absorbing the extraordinary array of new chemical compounds (many with highly toxic qualities) in biosystems ill prepared for them. Even though a broad scientific consensus exists concerning the potentially serious nature of such problems, the ability of scientists to predict impacts and outcomes of environmental transformations with accuracy is severely limited. Most willingly admit so. Their worries and concerns are guided as much by beliefs as by evidence. In the face of this unpredictability, the uncertainties, risks, and unintended consequences that attach to taking the wrong kind of pre-emptive action in the face of some imagined 'collision' may be just as bad as not taking action in the face of impending disaster. The problem, however, is that neither the environmental movement nor their allies within the scientific community are well prepared to acknowledge let alone take seriously the potentially negative consequences, for human beings as well as for other species, of the unintended consequences of their own proposals.

An unthinking crisis rhetoric also helps legitimize all manner of actions irrespective of social or political consequences. When the British Government came to the climate change conference in Kyoto in 1997 as one of the few countries that had met its target for the carbon dioxide reductions earlier proposed at the Rio Conference of 1994, it did so because of the Conservative Party's determination to crush the power of labor anchored in the Miners' Union by freeing the British energy industry from its dependency on coal. Closing the mines for class struggle reasons could be legitimized by appeal to environmental well-being. There are, unfortunately, far too many examples (historical as well as contemporary – see Harvey, 1996, Chapter 8) where social and political aims have been clothed or excused in the rhetoric of environmentalism, dictated by natural limits or nature-imposed scarcities. A rhetoric of impending environmental catastrophe will not, furthermore, necessarily sharpen our minds in the direction of cooperative, collective, and democratic responses. It often sparks elitist and authoritarian impulses (particularly among many scientists) or even a 'lifeboat ethic' in which the powerful pitch the rest overboard.

The invocation of 'limits' and 'ecoscarcity' should, therefore, make us as politically nervous as it makes us theoretically suspicious (see Harvey, 1996, 139–49). While there are versions of this argument that accept that 'limits' and 'ecoscarcities' are socially evaluated and produced (in which case the question of limits in nature gets so softened as to become almost irrelevant), it is hard to stop this line of thinking slipping into some version of naturalism (the absolutism of fixed limits in nature) or, worse still, a fatalistic Malthusianism in which disease, famine, war, and mutliple social disruptions are seen as the 'natural' correctives to human hubris. Not a few radical environmentalists now claim that Malthus was right rather than wrong.

Against the idea that we are headed over the cliff into some abyss (collapse) or that we are about to run into a solid and immovable brick wall (limits), I think it consistent with both the better sorts of environmental thinking and Marx's dialectical materialism to construe ourselves as embedded within an on-going flow of living processes that we can individually and collectively affect through our actions. We are profoundly affected by all manner of events (particularly physical changes in energy flows on earth, land, and sea, adaptations by other species as well as changes we ourselves induce). We are active agents caught within 'the web of life' (see, e.g., Capra, 1996, or Birch and Cobb, 1981, for elaborations of this concept). The dialectical conceptions laid out by Levins and Lewontin (1985) or the process-based arguments of someone like Whitehead (1969) can be brought together with those of Marx (see Harvey, 1996, Chapter 2) to consolidate the 'web of life' metaphor into a theoretical and logical structure of argument. I consider this by far the most useful metaphor to understand our situation. It is certainly more accurate and more useful than the linear thinking that has us heading off a cliff or crashing into a brick wall. It is therefore significant that the world scientists' warning to humanity shifts its metaphorical ground in mid-flight from that of a collision between two entities to the idea of 'the world's interdependent web of life' (Union of Concerned Scientists, 1996).

3 The web of life

The kind of nature we might be in a position to produce in years to come will have powerful effects upon emergent and even new social forms. How we produce nature in the here and now is therefore a crucial grounding for any dialectical utopianism. And how we construct the problem discursively also has its crucial moment in the sun as constituting the imaginative moment through which alternative visions can be constructed.

How, then, can perspectives on future alternatives be constructed from within the 'web of life' metaphor? We must first consider the directly 'negative' and 'positive' consequences of diverse past and present human activities, both for ourselves (with appropriate concern for class, social, national, and geographical distinctions) as well as for others (including non-human species and whole habitats). But even more importantly we need to recognize how our actions filter through the web of interconnections that make up the living world with all manner of unintended consequences. Like many other species, we are perfectly capable (without necessarily being conscious of it) of fouling our own nest or depleting our own resource base so as to seriously threaten the conditions of our own survival (at least in achieved cultural and economic if not in more basically physical terms). There are innumerable historical and geographical examples where human populations have, in effect, died in their own wastes and excrements (the original colonial settlements in Jamestown may have died out for this reason). But accepting this is different from thinking we are reaching some limit in nature, that environmental catastrophe is just around the corner, or, even more dramatically, that we are about to destroy the planet earth.

We can likewise overextend ourselves and create walls and limits around ourselves where there were none before. But these are self- and socially-created barriers and scarcities rather than nature-imposed limits and walls. This was, of course, the central point that George Perkins Marsh drove home with historical-geographical example after example in that remarkable book *Man and Nature* published back in 1864. He wrote:

> [I]t is certain that man has done much to mould the form of the earth's surface, though we cannot always distinguish between the results of his action and the effects of purely geological causes ... The physical revolutions thus wrought by man have not all been destructive to human interests ... [But] man has too long forgotten that the earth was given to him for usufruct alone, not for consumption, still less for profligate waste. Nature has provided against the absolute destruction of any of her elementary matter, the raw material of her works ... But man is everywhere a disturbing agent. Wherever he plants his foot, the harmonies of nature are turned to discords. The proportions and accommodations which insured the stability of existing arrangements are overthrown. Indigenous vegetable and animal species are extirpated, and supplanted by others of foreign origin, spontaneous production is forbidden and restricted, and the fate of the earth is either laid bare or covered with a new and reluctant growth of vegetable forms, and with alien tribes of animal life. These intentional changes and substitutions constitute, indeed, great revolutions; but vast as is their magnitude and important, they are, as we shall see, insignificant in comparison with the contingent and unsought results which have flowed from them.
>
> (Marsh, 1965 edition, 18)

Merely monitoring such interactions, threats, and environmental challenges turns out to be a huge and complicated task. It requires the deployment of massive research and intellectual resources within a hard-to-coordinate academic division of labor across a wide range of political and ideological beliefs (the work of the International Panel on Climate Change these last few years shows it can be done, at least in a fashion).

The issue of scale, both temporal and spatial, is here vital to how we identify and assess the seriousness of environmental issues and try to track unintended consequences. Global issues (warming and loss of biodiversity) contrast with micro-local issues (radon in the basement) and short-term difficulties intermingle with long-term trends. The theory of uneven geographical developments (see Chapter 5), with its emphasis upon scalars and differentiations, is as applicable here as anywhere. Relations between scales must be understood because our responsibilities to nature and to our species being comprise actions that vary from the micro preservation of habitat diversities in hedgerows and in nooks and crannies of gardens through regional issues like deteriorating water resources and tropospheric ozone concentrations to the hugely complicated global issues of strato-spheric ozone depletion, resource degradation, maintenance of biodiver-sity, and global warming. Dialectical utopianism has to incorporate such issues within its compass, for this is the ecological world we have to change as we seek to change ourselves.

How, then, should we generalize about our contemporary situation and the alternatives to which it points? The environmental movement backed by science has pioneered in alerting us to many of the risks and un-certainties to be confronted. There is far more to the environmental issue than the conventional Malthusian view that population growth might outstrip resources and generate crises of subsistence (up until as late as the 1970s this was the dominant form environmentalism took). A strong case can be made that the humanly-induced environmental transformations now under way are larger scale, riskier, and more far reaching and complex in their implications (materially, spiritually, aesthetically) than ever before in human history (as Marsh pointed out over a century ago and many others, such as Beck, 1992, have emphasized). The quantitative shifts that have occurred in the last half of the twentieth century in, for example, scientific knowledge and engineering capacities, industrial output, waste generation, invention of new chemical compounds, urbanization, popula-tion growth, international trade, fossil fuel consumption, resource extrac-tion, habitat modification – just to name some of the most important features – imply a qualitative shift in environmental impacts and potential unintended consequences that requires a comparable qualitative shift in our responses and our thinking. The evidence for widespread unintended consequences of the massive environmental changes now under way (some distinctly harmful to us and others unnecessarily harmful to other species), though not uncontested, is now persuasive (cf. loss of biodiversity at accelerating rates). The web of planetary life has become so permeated with human influences that evolutionary paths depend heavily (though by no means exclusively) on our collective activities and actions. For this reason alone it is important to side with caution. While I do not accept the apocalyptic rhetoric of limits and catastrophe as an overarching metaphor, I do not therefore dismiss all worrying data and all serious concerns as 'merely alarmist' as do many environmental sceptics.

In this regard, Wilson (1998, 290–2) has it roughly right. We face a series of environmental bottlenecks in the twenty-first century which have already largely been constructed by past human actions. It is important to take evasive action now to prevent them closing to form solid walls. I agree with his view that we not only have a responsibility to emerge from those bottlenecks in a 'better' condition than we entered (though I would define 'better' in a different way from him), but we also need to ensure that we take as much of the rest of life with us as possible.

Prudence in the face of such mounting risks is a perfectly reasonable posture. It also provides a more likely basis for forging some collective sense of how to exercise our responsibilities to nature as well as to human nature. But matters are far from simple. To begin with, the definition of 'environmental issues' often entails a particular bias, with those that affect the poor, the marginalized, and the working classes frequently being ignored (e.g. occupational safety and health) while those that affect the rich and the affluent get emphasized (e.g. poverty is a far more important cause of shortened life expectations in the United States than smoking but it is smoking that gets all the attention). Secondly, environmental impacts frequently have a social bias (class, racial, and gender discriminations are evident in, say, the location of toxic waste sites and the global impacts of resource depletion or environmental degradation). Thirdly, some risks and uncertainties can strike anywhere, even against the rich and the powerful. The smoke from the fires that raged in Indonesia in the Fall of 1997 did not respect national or class boundaries any more than did the cholera that swept nineteenth-century cities. Problems of the latter sort can sometimes provoke a universal, rather than a specifically class-based, approach to public health and environmental regulation. The threat of increased hurricane frequencies from global warming terrifies insurance companies as much as it irritates the auto and oil companies to hear that they should cut back on their global plans for expansion because of the threat of emissions to the atmosphere (though even here the drive to produce a non-polluting car is becoming more and more evident within the auto industry itself). Finally, the distinction between the production/ prevention of risks and the capitalistic bias towards consumption/com-modification of cures has significance.

The implication is that there are multiple contradictions to be worked out as we contemplate responsibilities to nature on the one hand and responsibilities to human nature on the other. The latter are not by definition antagonistic to the former. But traditional ways of looking upon the solution to poverty as lying entirely with redistributions out of growth (or, for that matter, that the path towards communism lies solely by way of liberation of the productive forces) cannot easily be sustained. Other ways of achieving such social and political objectives must be found. And there are now, fortunately, abundant examples of how more equitable access to life chances can produce environmentally beneficial as well as socially advantageous results (cf. the case of Kerala).

Environmental arguments are not necessarily or even broadly antag-onistic to class politics. An evolving socialist or other alternative perspec-tive needs to understand the specific class content and definition of environmental issues and seek alliances around their resolution (as, for example, in the environmental justice movement). Furthermore, many issues that start out as non-class can end up having a strong class content as remedies are sought and applied. For example, AIDS and poverty are becoming increasingly interlocked because the expensive ways of control-ling the spread of AIDS are leaving poor countries in Africa or impover-ished inner-city populations most horribly exposed (in some of the poorest countries of Africa nearly a quarter of the population is HIV positive). The politics of any class-based environmental movement entails building alliances across many social layers in the population including many of those not directly affected by the issue at hand.

But there is a more general point. The risk and uncertainty we now experience acquires its scale, complexity, and far-reaching implications by virtue of processes that have produced the massive industrial, technolo-gical, urban, demographic, lifestyle, and intellectual transformations and uneven developments that we have witnessed in the latter half of the twentieth century. In this, a relatively small number of key institutions, such as the modern state and its adjuncts (including international co-ordinating agencies), multinational firms, and finance capital, and 'big' science and technology, has played a dominant and guiding role. For all the inner diversity, some sort of hegemonic economistic-engineering discourse has also come to dominate discussion of environmental ques-tions and to dictate how we are to construe our reponsibilities to nature and to human nature. Commodifying everything and subjecting almost all transactions (including those connected to the production of knowledge) to the singular logic of commercial profitability and the cost-benefit calculus is a dominant way of thinking. The production of our environ-mental difficulties, both for the working class, the marginalized, and the impoverished (many of whom have had their resource base stripped from under them by a rapacious commercialism) as well as for some segments of capital and even of some elements of the rich and the affluent, is consequential upon this hegemonic class project, its market-based philo-sophy and modes of thinking that attach thereto.

This invites as response the organization of an equally powerful class project of risk prevention and reduction, resource recuperation and control, in which the working class, the disempowered, and the marginalized take a leading role. In performing that role the whole question of constructing an alternative mode of production, exchange, and consumption that is risk reducing and environmentally as well as socially just and sensitive can be posed. Such a politics must rest on the creation of class alliances – including disaffected scientists who see the problem but have little conception of how to construct a socially just solution – in which the environmental issue and a more satisfying 'relation to nature' have a prominent place alongside the reconstruction of social relations and modes of production and consumption. A political project of this sort does not, I insist, need a rhetoric of limits or collapse to work effectively and well. But it does require careful and respectful negotiation with many environmental movements and disaffected scientists who clearly see that the way contemporary society is working is incompatible with a satisfactory resolution to the environmental questions that so bother them. The basis for such a project must rest, however, upon some broad agreement on how we are both individually and collectively going to construct and exercise our responsibilities to nature in general and towards our own human nature in particular.

4 Learning to be distinctively ourselves in a world of others

The construction of some broad political movement around the whole issue of 'responsibility to nature and to human nature' requires negotiation and translation between diverse habits of mind that derive from the uneven ways in which material life, social practices, and knowledge systems are orchestrated and organized. We are necessarily anthropocentric, ethnocentric, and auto-centric. Nevertheless, even if our task is, as White (1990, 257–64 [264]) puts it, 'to be distinctively ourselves in a world of others,' then there are still various ways in which we can 'be ourselves.' The choices we make and the practices we engage in have everything to do with constructing our species potential.

The twentieth century has in many respects been dominated by introspective approaches to such questions. Much of the great art and literature of our time, and not a little of the academic enquiry prosecuted in the last thirty years in terms of understanding situated knowledges, positionalities, and the like has been dominated by the quest to understand the inner self (either of oneself or, with the aid of psychoanalytic findings, of others) and to understand forms of expression, texts, representations, and even whole symbolic systems in terms of their inner meanings. Such deconstructive and introspective techniques have yielded much of value but have the habit, particularly when set up in a relational or dialectical mode, of turning inside out. The quest to understand inner meanings is inevitably connected to the need to understand relations with others. Arne Naess, a deep ecologist, argues, for example, that it is essential to replace the narrow and ineffective egotistical concept of 'self' with the broader idea of 'Self' as an internal relation of all other elements in the natural world. Derrida claims that the only true form of expression is that which unchains the voice of the internalized other within ourselves.

There is, therefore, an external frontier of understanding which still remains relatively untilled and unexplored. This concerns relations with others who live their lives and have their being in different material circumstances and whose experiences are shaped in different sensory worlds to those which we immediately experience. In this regard, we can, if we wish, 'create a frame that includes both self and other, neither dominant, in an image of fundamental equality' (White, 1990, 264). We can strive to construct our actions in response to a wide range of imaginaries, even if it is still we who do the thinking and we who choose to use our capacities and powers *this* way rather than *that*. Such a principle applies to all 'others' be it the 'grand other' of 'nature' or 'others' within our own species who have pursued different imaginaries with different under- standings born out of different cultural configurations and material prac- tices. In the latter case, however, my capacity to empathize and 'think like the other' is further aided by the possibility to translate across the languages and the varied discursive regimes (including all manner of systems of representation) that express often radically different attitudes to how we should and actually do, both individually and collectively, construct ourselves through transformations of the world.

So though it is always an 'I' or a 'we' who does the imagining and the translation, and although it is always in the end through my (our) language that thinking gets expressed, there are many different ways in which we can hope to build frames of thought and action that relate across self and others. I can try to 'think like' a mountain or a river or a spotted owl or even as the ebola virus in ways that many ecologists urge and in so doing regulate my ways of thinking and being in the world differently from those who prefer to 'think like' Rupert Murdoch or Australian aborigines. We exercise such choices in part because that is how we can explore our capacities and powers and become something other than what we already are. If respect and love of others is vital to respect and love of self, then we should surely approach all others, including those contained in what we often refer to as 'external nature,' in exactly such a spirit. Concern for our environment is concern for ourselves.

Two compelling conclusions then follow. First, that the long-lost techniques of empathy and translation (see Chapter 12) across sensory realms is reconstituted as a vital way of knowing to supplement (and in certain instances to transcend) introspection and all the various objectivizing modes of enquiry that exist mainly in the sciences and social sciences. Secondly, 'where and who we learn it from and how we learn it' overrides the contemporary postmodern fascination with 'where we see it from' as the basis for intellectual engagements. Knowledges are and can be constituted in a variety of ways and how they are constructed plays a crucial role in our ability to interpret and understand our way of being in the world.

5 Thoughts on the unity of knowledge

From time to time the idea has arisen that there is (or, more often, there ought to be) some unity within the evidently diverse knowledges human beings possess. In recent times the search for such a unity has been all but abandoned within the humanities and even most of the social sciences. All 'totalizing systems of thought' have been judged wanting and therefore beyond the pale of discussion. The effect has been to leave the whole question of the potential unity of knowledges to mavericks, religious fanatics, and a small select band of thinkers – mainly drawn from the natural sciences – still admiring of enlightenment traditions.

It has not always been so. The logical positivists of the Vienna Circle mounted a serious quest for the unity of knowledge through the examination of symbolic and mathematical forms from the 1930s onwards. But this effort faded in the 1960s largely as a consequence of the movement's manifest failure to unlock an adequate key to the problem of how language in itself could uniquely represent the world. Wittgenstein persuasively showed that when languages are understood as language games they lose their pre-eminent role as the privileged site where unification might be found even if, as Chomsky has long contentiously argued, capacities for a deep structure of language are inherent to our species being.

The Marxist version of the unity of knowledge, mainly set out in the early works such as *The Economic and Philosophic Manuscripts* of 1844, likewise fell apart because its increasingly formalistic interpretation (given a hearty push by Lenin among others) made it too internally contradictory for its own good. The internal contradictions ended up being expertly exposed by, among others, Althusser. That current of thought dubbed 'Althusserian' within Marxism deconstructed the notion of any simple formalistic unity (based on the idea that material circumstances determine states of consciousness, for example) but failed to put anything else substantial in its place. The frontal assault upon all forms of 'metanarrative' from the mid-1970s onwards spelled doom for Marxist ambitions when coupled with this deep internal questioning within the Marxist tradition itself.

The effect has been to leave serious debate on the question of unity to the scientists. The complex problem of the unity of diverse *knowledges* has thereby been reduced to the more specific quest for unity within the *sciences*. Much can be learned from these recent efforts – from general systems theory, complexity theory, or even the vaguer but intriguing arguments that surround the notions of Gaia. But such efforts only make sense when translated back into the quest for a unity of all manner of knowledges, including those that are of a non-scientific nature.

Consider, for example, the arguments put forward by E. O. Wilson (1998). He insists on the unity of knowledge as a primary goal in liberating the human imaginary to confront the responsibilities we now have to both nature and human nature. Such a unity – which he calls 'consilience' – crucially depends upon an ability to work at and relate across different genres (from physics to aesthetics and ethics) and scales. With respect to the latter he argues that:

> [The] conception of scale is the means by which the biological sciences have become consilient during the past fifty years. According to the magnitude of time and space adopted for analysis, the basic divisions of biology are from top to bottom as follows: evolutionary biology, ecology, organismic biology, cellular biology, molecular biology, and biochemistry ... The degree of consilience can be measured by the degree to which the principles of each division can be telescoped into those of the others.   (Wilson, 1998, 83)

The question of scale, I argued in Chapter 5, is fundamental to any theory of uneven geographical development. So I listen carefully to Wilson on this point. The evident successes of the biological sciences as a whole, he argues, have depended crucially on a readiness to embed understandings at one scale (e.g. evolution) in findings generated at another (e.g. genetics) thus establishing consilience across spatiotemporal scales. Although this success is threatened by an increasingly myopic approach on the part of individual scientists working within the divisions of scientific labor, 'the focus of the natural sciences has begun to shift away from the search for new fundamental laws and toward new kinds of synthesis – "holism" – if you prefer – in order to understand complex systems' (267).

Consilience depends, however, on general acceptance of a scientific method that brings the disciplines together even in the absence of the will of individual researchers. 'The strategy that works best,' he writes, 'is the construction of coherent cause-and-effect explanations across levels of organization' (267). This reductionism forms the central guiding thread (perhaps better understood as a coercive disciplinary apparatus of rules of engagement) for constructing the unity not only of the sciences but of all other forms of knowledge:

> The central idea of the consilience world view is that all tangible phenom-
> ena, from the birth of stars to the workings of social institutions, are based
> on material processes that are ultimately reducible, however long and tor-
> tuous the sequences, to the laws of physics.   (266)

Such a view will gain little sympathy in the humanities or the social sciences (particularly given the provocative way Wilson characterizes knowledges in these fields). Nor will his arguments garner much sympathy on the left (except, perhaps, among ecologists) because of the fear of any resort to biologism, naturalism, or reductionism. There are, however, all sorts of interesting internal tensions in Wilson's argument. He never does solve the problem of how a reductionism to physical law can explain the rise of civilization or even the construction of knowledge (calculus and Mozart remain unexplained). He resorts to the pure *belief* (without any scientific evidence whatsoever) that reductionism can work 'in principle' even if the practical possibility is remote. He recognizes vaguely that the task of synthesis requires a different kind of intelligence and procedure from the mere additive cause-and-effect reductionism he formally advo-cates. And as evolution becomes volitional he ends up arguing that we should 'look deep inside ourselves' as to what we want to become, thereby inverting the causal ordering from physics and evolutionary biology to moral and ethical choice. In the end it is the humanization of science that gets advocated, though his reductionism overtly points in precisely the opposite direction.

Marx, it is useful to note, was not averse to similarly reductionist arguments. He rooted his thinking in Darwin's theory of evolution (while criticizing Darwin's actual formulations):

> Darwin has directed attention to the history of natural technology, i.e. the
> formation of the organs of plants and animals, which serve as the instru-
> ments of production for sustaining their life. Does not the history of the
> productive organs of man in society, of organs that are the material basis of
> every particular organization of society, deserve equal attention? And would
> not such a history be easier to compile since, as Vico says, human history
> differs from natural history in that we have made the former, but not the
> latter? Technology reveals the active relation of man to nature, the direct
> process of production of his life, and thereby it also lays bare the process of
> production of the social relations of his life, and of the mental conceptions
> that flow from those relations. Even a history of religion that is written in

abstraction from this material basis is uncritical. It is, in reality, much easier
to discover by analysis the earthly kernel of the misty creations of religion
than to do the opposite, i.e. to develop from the actual, given relations of life
the forms in which these have been apotheosized. The latter method is the
only materialist, and therefore the only scientific, one. The weaknesses of
the abstract materialism of natural science, a materialism which excludes
the historical process, are immediately evident from the abstract and
ideological conceptions expressed by its spokesmen whenever they venture
beyond the conceptions of their own speciality.

(Marx, 1976 edition, 493–4)

It would be intriguing to get Wilson's commentary on such a passage. But
several points can usefully be made about this, one of Marx's most
considered methodological statements. To begin with, theorizing is here
left open to accommodations to the evolutionary process as well as to
advances in evolutionary science. Marx does see a qualitative shift as
history differentiates itself from biological evolution. This qualitative shift
is construed dialectically and not in causative mechanistic terms of the sort
that Wilson prefers. Furthermore, it does not entail any radical break with
evolution but effectively grafts an understanding of human endeavors and
of the historical-geography of human activity into the web of evolutionary
change.

The difficulty with Wilson's vision (leaving aside its internal incon-
sistencies, unjustifiable claims, and polemical short cuts) is that he cannot
openly admit (except as an opening and closing gambit about his own
beliefs) in the historical and humanizing side of what he must necessarily
profess in his quest for some unity of science. This, however, is typical of
many of the past attempts to uncover the unity of knowledge. It presumes
that such unity depends on a single thread (hence the thirst for reductionism) and that the end point is stability and harmony of completely
achieved understanding. For Wilson the single thread is a causative-
mechanistic principle which provides reductionist links and the end point
lies where everything (including ethics and aesthetics) is brought within its
domain as a harmonious whole. This is, however, just one possible (and
inherently limited) mode of unification that can operate successfully only
under limited conditions and within a limited domain (let us grant, for
example, that it has worked in the way Wilson describes in some of the
biological sciences).

Against this predominant mode of thought, I suggest we view knowledges as constructed more on the model of complex interlocking ecologies
made up of distinctive processes and parts dynamically feeding off each
other in often confusing and contradictory ways. Within this system it is
possible to think of 'families of meanings' (a phrase that Wittgenstein

favored as opposed to the 'mutually exclusive meanings' sometimes inferred from his theory of language games). Like any extended family, there are abundant interactions, interdependencies, differences, and not a little contentiousness and conflict (including, on occasion, violent and internecine struggles between different clans and branches). This is exactly the case with the extraordinary diversity of discourses about the environment.

Wilson, interestingly, ends up constructing an example of exactly such an interdigitation of knowledge structures. He sets up an irresolvable but dynamic conversation between a 'transcendentalist' and 'empiricist' world view, a conversation between two different modes of knowing. But he views them as mutually exclusive rather than internally related. Wilson's own history as a Christian and a rigorous scientist – a history which he readily sees as both contradictory and enriching – plays into his theorizing at all levels. He here illustrates how radically different meanings can and must relate. Without faith his whole mission of scientific consilience would be inconceivable and without commonalities conversation would be impossible.

An ecological and evolutionary view of knowledge provides a sense of how to think about both the unity and the diversity of knowledges. The metaphor of the family, when extended through the idea of 'the family of man' to our 'species being' provides a way of thinking through the possible interrelations. Armed with a wide variety of capacities and powers – including those of empathy and translation as well as objectivized observations – it is possible to think of ways in which different families of meanings can be (internally) related to each other. The methodological stance becomes one of dialectics (in which, as Levins and Lewontin [1985, 278] put it, there is 'no basement' and within which it is therefore entirely legitimate to search for distinctive fundamental units and processes operating at distinctive scales and levels) rather than of a reductionist causation. Setting boundaries with respect to space, time, scale, and environment then becomes a major strategic consideration in the development of concepts, abstractions, and theories. It is usually the case that any substantial change in these boundaries will radically change the nature of the concepts, abstractions, and theories. Levins and Lewontin make a similar point with respect to time and change:

> The dialectical view insists that persistence and equilibrium are not the natural state of things but require explanation, which must be sought in the actions of the opposing forces. The conditions under which the opposing forces balance and the system as a whole is in stable equilibrium are quite special. They require the simultaneous satisfaction of as many mathematical relations as there are variables in the system, usually expressed as inequalities among the parameters of that system.   (1985, 275)

Nature, says Whitehead (1969, 33), is always about the perpetual exploration of novelty. And human nature is in this regard no exception.

Dialectical enquiry necessarily incorporates, therefore, the building of ethical, moral, and political choices (*values*) into its own process and sees the constructed knowledges that result as discourses situated in a play of power directed towards some goal or other. Values and goals (what we might call the 'teleological' as well as the 'utopian' moment of reflexive thought) are not imposed as universal abstractions from outside but arrived at through a living process (including intellectual enquiry) embedded in forms of praxis and plays of power attaching to the exploration of this or that potentiality (in ourselves as well as in the world we inhabit).

The unity of knowledges is, moreover, not seen in terms of harmony but in terms of an evolutionary process in which the facets of competition, diversification, collaboration, dispersal and diffusion, modifications of the objective world (both social and natural), and determinations of spatio-temporal orderings are all brought into play. This reveals perhaps the most striking paradox of all in Wilson's own presentation: that the only place where such evolutionary processes fail to play out their role is in the production of the consilient knowledge system itself.

6 Spatiotemporal utopianism and ecological qualities

Let us return to the figure of the architect. What kind of unity of knowledge and action is presupposed in what architects do? Karatani's (1995, XXXVIII) characterization of architectural practices is here helpful:

> Design is similar to Wittgenstein's term 'game,' where, as he says, 'we play – and make up the rules as we go along.' No architect can predict the result. No architecture is free of its context. Architecture is an event par excellence in the sense that it is a making or a becoming that exceeds the maker's control. Plato admired the architect as metaphor but despised the architect as an earthly laborer, because the actual architect, and even architecture itself, are exposed to contingency. Contingency does not imply, however, that, as opposed to the designer's ideal, the actual architecture is secondary and constantly in danger of collapse. Rather, contingency insures that no architect is able to determine a design free from the relationship with the 'other' – the client, staff, and other factors relevant to the design process. All architects face this other. Architecture is thus a form of communication conditioned to occur without common rules – it is a communication with the other, who, by definition, does not follow the same set of rules.

The architecture of dialectical utopianism must be grounded in contingent matrices of existing and already achieved social relations. These comprise political-economic processes, assemblages of technological capacities, and the superstructural features of law, knowledge, political beliefs, and the like. It must also acknowledge its embeddedness in a physical and ecological world which is always changing. To paraphrase Marx, we architects all exercise the will to create but do so under conditions not chosen or created by ourselves. Furthermore, since we can never be entirely sure of the full implications of our actions, the resultant trajectories of historical-geographical change always escape from the total control of our individual or collective wills.

This conception is antagonistic to a strongly binary tradition in Western thought that goes back to the Greeks. Karatani (1995, 5) summarizes it thus:

> On the one hand, evolutionists consider the world a living, growing form or organism; on the other hand, creationists consider it a designed work of art. These two types represent two worldviews: one that understands the world as becoming and another that understands the world as a product of making.

The contemporary version of this debate pits 'social constructivism' against the 'objective science' of, for example, genetic determination. But Marx does not fit easily into such a binary. Nor, interestingly, does Wilson in practice in spite of all his avowals to the contrary. Such a binary cannot, I conclude, capture what evolution in general and human evolution in particular might be about. There is, as Marx insisted, nothing 'unnatural' about the historical geography of human development. We act upon the world as a 'force of nature' and, like all architects, seek to create works of art whose implications we can never fully understand or control.

The challenge is, therefore, to work out a language for dialectical utopianism that is materially grounded in social and ecological conditions but which nevertheless emphasizes possibilities and alternatives for human action through the will to create. The ecological dimension to utopian thinking has, of course, its own tradition. But it has either been marginalized (most classical utopian schemes say little or nothing about the negotiation with nature and if they do it tends to presume a land like Cockaigne, full of milk and honey) or far too restrictive, usually predicated upon some doctrine of the harmony with nature achieved at some relatively small scale (the commune, the village, or the small town). Many residues of a utopian environmentalism can be detected in the landscape of capitalism (garden cities, suburbs, and access to recreational areas as well as movements for clean air, clean water, and adequate sewage disposal in metropolitan regions). While such achievements and movements should not be discounted, we have plainly gone beyond situations easily addressed by such modes of thought. Our collective responsibilities to human nature and to nature need to be connected in a far more dynamic and co-evolutionary way across a variety of spatiotemporal scales. Issues like conservation of micro habitats, ecological restoration projects, urban design, fossil fuel utilization, resource exploitation patterns, livelihood protections, sustenance of certain geographically specific cultural forms, enhancement of life chances at everything from the global to the local level, all somehow need to be brought together and factored into a more generalized sense of how a political-economic alternative might arise out of the ecological contradictions of a class-bound capitalist system.

We can all seek to be architects of our fates by exercising our will to create. But no architect is ever exempt from the contingencies and constraints of existing conditions and no architect can ever hope, except in that realm of pure fantasy that does not matter, to so control the web of life as to be free of 'the contingent and unsought results' which flow from their actions. Architects and bees at least have that in common, even if what distinguishes them also clearly signals where and how the real political movement to abolish the present abysmal state of things can be set in motion.

CHAPTER 12

The insurgent architect at work

Imagine ourselves as architects, all armed with a wide range of capacities and powers, embedded in a physical and social world full of manifest constraints and limitations. Imagine also that we are striving to change that world. As crafty architects bent on insurgency we have to think strategically and tactically about what to change and where, about how to change what and with what tools. But we also have somehow to continue to live in this world. This is the fundamental dilemma that faces everyone interested in progressive change.

But what kind of world are we embedded in? We know that it is a world full of contradictions, of multiple positionalities, of necessary flights of the imagination translated into diverse fields of action, of uneven geographical developments, and of highly contested meanings and aspirations. The sheer enormity of that world and its incredible complexity provide abundant opportunities for the exercise of critical judgement and of limited freedom of the individual and collective will. But the enormity of apparent choice and the divergent terrains upon which struggles can be conducted is perpetually in danger of generating a disempowering confusion (of the sort that globalization, for example, has strongly promoted). Furthermore, it appears impossible to avoid unintended consequences of our actions, however well thought out. How are we to cut through these confusions and build a different sense of possibilities while acknowledging the power of the constraints with which we are surrounded?

Here are some conversation points in lieu of answers. In the last chapter I argued for a system of *translations* across and between qualitatively different but related areas of social and ecological life. The spatiotemporal scale at which processes operate here makes a difference. For this reason, Wilson considers scale as one of the most important differentiations within the unity of science. *The Communist Manifesto* notes the same problem as revolutionary sentiment passes from the political individual through the factory, political parties, and the nation state to a movement in which workers of the world can unite. Dialectics permits diverse

233

knowledges and practices to be rendered coherent across scales without resort to some narrow causal reductionism. This dialectical way of thinking echoes aspects of the theory of uneven geographical develop-ments laid out in Chapter 5. I there suggested that the production of spatiotemporal scale is just as important as the production of differentia-tions within a scale in defining how our world is working and how it might work better.

I now take up these ideas in greater depth. I propose first that we consider political possibilities at a variety of spatiotemporal scales. I then argue that real political change arises out of simultaneous and loosely coordinated shifts in both thinking and action across several scales (either simultaneously or sequentially). If, therefore, I separate out one particular spatiotemporal scale for consideration, in order to understand its role in the overall dynamics of political change, then I must do so in a way that acknowledges its relation to processes only identifiable at other scales. The metaphor to which I appeal is one of several different 'theaters' of thought and action on some 'long frontier' of 'insurgent' political prac-tices. Advances in one theater get ultimately stymied or even rolled back unless supported by advances elsewhere. No one theater is particularly privileged even though some of us may be more able, expert, and suited to act in one rather than another. A typical political mistake is the thoroughly understandable habit of thinking that the only theater that matters is the one that I or you happen to be in. Insurgent political practices must occur in all theaters on this long frontier. A generalized insurgency that changes the shape and direction of social life requires collaborative and coordinat-ing actions in all of them. With that caveat in mind, I consider seven theaters of insurgent activity in which human beings can think and act, though in radically different ways, as architects of their individual and collective fates.

1 The personal is political

The insurgent architect, like everyone else, is an embodied person. That person, again like everyone else, occupies an exclusive space for a certain time (the spatiotemporality of a human life is fundamental). The person is endowed with certain powers and skills that can be used to change the world. He or she is also a bundle of emotions, desires, concerns, and fears all of which play out through social activities and actions. The insurgent architect cannot deny the consequences of that embodiment in material, mental, and social life.

Through changing our world we change ourselves. How, then, can any of us talk about social change without at the same time being prepared, both mentally and physically, to change ourselves? Conversely, how can we change ourselves without changing our world? That relation is not easy to negotiate. Foucault (1984) rightly worried that the 'fascism that reigns in our heads' is far more insidious than anything that gets constructed outside.

Yet we also have to decide – to build the road, the factory, the houses, the leisure park, the wall, the open space . . . And when a decision is made, it forecloses on other possibilities, at least for a time. Decisions carry their own determinations, their own closures, their own authoritarian freight. Praxis is about confronting the dialectic in its 'either/or' rather than its transcendent 'both/and' form. It always has its existential moments. Many of the great architects of the past made their personal political in incredibly decisive as well as authoritarian ways (with results both good and bad according to the partial judgement of subsequent generations).

It is in this sense, therefore, that the personal (including that of the architect) is deeply political. But that does not mean, as feminists, ecologists, and the innumerable array of identity politicians who have strutted their stuff these last few years have discovered to their cost, that virtually *anything* personal makes for good politics. Nor does it mean, as is often suggested in some radical alternative movements (such as deep ecology), that fundamental transformations in personal attitudes and behaviors are sufficient (rather than necessary) for social change to occur.

While social change begins and ends with the personal, therefore, there is much more at stake here than individualized personal growth (a topic that now warrants a separate and large section in many bookstores in the United States) or manifestations of personal commitment. Even when it seems as if some charismatic and all-powerful person – a Haussmann, a Robert Moses, or an Oscar Niemeyer – builds a world with the aim of shaping others to conform to their particular and personal visions and desires, there turns out to be much more to it than just the vision of the person. Class interests, political powers, the mobilization of forces of violence, the orchestration of discourses and public opinion, and the like, are all involved.

But in reflecting on what we insurgent architects do, a space must be left for the private and the personal – a space in which doubt, anger, anxiety, and despair as well as certitude, altruism, hope, and elation may flourish. The insurgent architect cannot, in the end, suppress or repress the personal any more than anyone else can. No one can hope to change the world without changing themselves. The negotiation that always lies at the basis of all architectural and political practices is, therefore, between persons seeking to change each other and the world, as well as themselves.

2 The political person is a social construct

To insist on the personal as political is to confront the question of the person and the body as the irreducible moment (defined at a particular spatiotemporal scale) for the grounding of all politics and social action. But the individual, the body, the self, the person (or whatever term we wish to use) is a fluid social construct (see Chapters 6 and 7) rather than some absolute and immutable entity fixed in concrete. How 'social construction' and 'embodiment' is understood then becomes important. For example, a relational conception of self puts the emphasis upon our porosity in relation to the world of socio-ecological change and thereby tempers many theories of individual rights, legal status, and the like. The person that is political is then understood as an entity open to the innumerable processes (occurring at different spatiotemporal scales) that transect our physical and social worlds. The person must then be viewed as an ensemble of socio-ecological relations.

But an already-achieved spatiotemporal order can hold us to some degree apart from this fluid and open conception in our thought and practices. In the United States, private property and inheritance, market exchange, commodification and monetization, the organization of economic security and social power, all place a premium upon personalized private property vested in the self (understood as a bounded entity, a non-porous individual), as well as in house, land, money, means of production, etc., all construed as the elemental socio-spatial forms of political-economic life. The organization of production and consumption forges divisions of labor and of function upon us and constructs professionalized personas (the architect, the professor, and the poet as well as the proletarian, all of whom, as Marx and Engels point out in *The Communist Manifesto*, 'have lost their halo' and become in some way or another paid agents of bourgeois power). We live in a social world that converts all of us into fragments of people with particular attachments, skills, and abilities integrated into those powerful and dynamic structures that we call a 'mode of production.' Our 'positionality' or 'situatedness' in relation to that is a social construct in exactly the same way that the mode of production is a social creation. This 'positionality' defines who or what we are (at least for now). And 'where we see it from' within that process provides much of the grist for our consciousness and our imaginary.

But 'what and how far we can see' from 'where we see it from' also varies according to the spatiotemporal constructions and our choices in the world we inhabit. Access to information via the media, for example, and the qualities and controls on information flow play an important role in how we can hope to understand and change the world. These horizons, both spatial and temporal, have simultaneously expanded and compressed over the past thirty years and part of any political project must be to intervene in the resultant information flows in ways that are progressive and constructive. But there is also the need to persuade people to look beyond the borders of that myopic world of daily life that we all necessarily inhabit.

In contrast, the fierce spatiotemporalities of daily life – driven by technologies that emphasize speed and rapid reductions in the friction of distance and of turnover times – preclude time to imagine or construct alternatives other than those forced unthinkingly upon us as we rush to perform our respective professional roles in the name of technological progress and endless capital accumulation. The material organization of production, exchange, and consumption rests on and reinforces specific notions of rights and obligations and affects our feelings of alienation and of subordination, our conceptions of power and powerlessness. Even seemingly new avenues for self-expression (multiculturalism being a prime recent example) are captive to the forces of capital accumulation (e.g. love of nature is made to equal eco-tourism). The net effect is to limit our vision of the possible. No less a person than Adam Smith (cited in Marx, 1976 edition, 483) considered that 'the understandings of the greater part of men are necessarily formed by their ordinary employments' and that 'the uniformity of (the labourer's) life naturally corrupts the courage of his mind.' If this is only partially true – as I am sure it is – it highlights how the struggle to think alternatives – to think and act differently – inevitably runs up against the circumstances of and the consciousness that derives from a localized daily life. Most insidious of all, is the way in which routine, by virtue of its comfort and security, can mask the ways in which the jarring prospects of transformative change must in the long run be confronted. Where, then, is the courage of our minds to come from?

Let us go back to the figure of the insurgent architect. She or he acts out a socially constructed (sometimes even performative) role, while confronting the circumstances and consciousness that derives from a daily life where demands are made upon time, where social expectations exist, where skills are acquired and supposed to be put to use in limited ways for purposes usually defined by others. The architect then appears as a cog in the wheel of capitalist urbanization, as much constructed by as constructor of that process (was this not as true of Haussmann, Cerda, Ebenezer Howard, Le Corbusier, Oscar Niemeyer, as of everyone else?).

Yet the architect can (indeed must) desire, think and dream of difference. And in addition to the speculative imagination which he or she necessarily deploys, she or he has available some special resources for critique, resources from which to generate alternative visions as to what might be possible. One such resource lies in the tradition of utopian thinking. 'Where we learn it from' may then become just as, if not more, important as 'what we can see from where we see it from.'

Utopian schemas of spatial form typically open up the construction of the political person to critique. They do so by imagining entirely different systems of property rights, living and working arrangements, all manifest as entirely different spatial forms and temporal rhythms. This proposed re-organization (including its social relations, forms of reproductive work, its technologies, its forms of social provision) makes possible a radically different consciousness (of social relations, gender relations, of the relation to nature, as the case may be) together with the expression of different rights, duties, and obligations founded upon collective ways of living.

Postulating such alternatives allows us to conduct a 'thought experiment' in which we imagine how it is to be (and think) in a different situation. It says that by changing our situatedness (materially or mentally) we can change our vision of the world. But it also tells us how hard the practical work will be to get from where we are to some other situation like that. The chicken-and-egg problem of how to change ourselves through changing our world must be set slowly but persistently in motion. But it is now understood as a project to alter the forces that construct the political person, my political person. I, as a political person, can change my politics by changing my positionality and shifting my spatiotemporal horizon. I can also change my politics in response to changes in the world out there. None of this can occur through some radical revolutionary break (though traumatic events and social breakdowns have often opened a path to radically different conceptions). The perspective of a long revolution is necessary.

But to construct that revolution some sort of collectivization of the impulse and desire for change is necessary. No one can go it very far alone. But positioned as an insurgent architect, armed with a variety of resources and desires, some derived directly from the utopian tradition, I can aspire to be a subversive agent, a fifth columnist inside of the system, with one foot firmly planted in some alternative camp.

3 The politics of collectivities

Collective politics are everywhere but they usually flow in constrained and predictable channels. If there is any broad swathe of insurgent politics at work in the interstices of urbanization in the advanced capitalist countries, for example, it is a mobilization in defense of private property rights. The violence and anger that greets any threat to those rights and values – be it from the state or even from agents of capital accumulation like developers – is an awesome political force. But it typically turns inwards to protect already existing personalized 'privatopias.' The same force can be found in the militia or neo-fascist movements on the right (a fascinating form of insurgent politics) as well as within the radical communitarianism of some ecologists.

Such formations of collective governance preclude the search for any far-reaching alternatives. Most politics and collective forms of action preserve and sustain the existing system, even as they deepen some of its internal contradictions, ecologically, politically, and economically (e.g. the collective rush to suburbanize increases car dependency, generates greenhouse gasses, particulate matter pollution, and tropospheric ozone concentrations etc.). The gated communities of Baltimore are a symbol of collective politics, willingly arrived at, gone awry.

Traditional utopianism seeks to confront this prevailing condition. Communitarianism as a utopian movement typically gives precedence, for example, to citizenship, to collective identifications and responsibilities, over the private pursuit of individual advantage and the 'rights talk' that attaches thereto. This ideal founds many a utopian dream, from Thomas More to Fourier, and infuses many contemporary religious movements like those for a Christian Base Community or even the much softer (and some would say much weaker) cultivation of concepts of 'citizenship' as the basis for the good life (see, e.g., Douglass and Friedmann, 1998).

Distinctive communities are painstakingly built by social practices including the exercise of authoritarian powers and conformist restrictions. They are not just imagined (however important the imaginary of them may be). It is useful, therefore, to view an achieved 'community' as an enclosed space (irrespective of scale or even frontier definitions) within which a certain well-defined system of rules prevails. To enter into that space is to enter into a space of rules which one acknowledges, respects, and obeys (either voluntarily or through some sort of compulsion). The construction of 'community' entails the production of such a space. Challenging the rules of community means challenging the very existence of such a collectivity by challenging its rules. It then follows that communities are rarely stable for long. Abundant opportunities exist here for the insurgent architect to promote new rules and/or to shape new spaces. Our capacities as rule makers and rule breakers here enters fully into play. Part of the attraction of the spatial form utopian tradition is precisely the way in which it creates an imaginary space in which completely different rules can be contemplated. And it is interesting to note how the figure of the city periodically re-emerges in political theory as the spatial scale at which ideas and ideals about democracy and belonging can best be articulated.

It is not always easy here to define the difference between insurgent politics of a progressive sort and the exclusionary and authoritarian practices of, say, homeowner associations who defend their property rights. Etzioni (1997), a leading proponent of the new communitarianism, actively supports, for example, the principle of closed and gated communities as a progressive contribution to the organization of social life. Collective institutions can also end up merely improving the competitive strength of territories in the high stakes game of the uneven geographical development of capitalism (see, e.g., Putnam's 1993 account of the institutional bases of uneven geographical development in Italy). For the privileged, community often means securing and enhancing privileges already gained. For the underprivileged it all too often means 'controlling their own slum.'

Dialectical utopianism must confront the production of 'community' and 'coming together for purposes of collective action' in some fashion and articulate the place and meaning of this phenomena within a broader frame of politics. This means a translation to a different scale from that of the embodied political person. Community should be viewed as a delicate relation between fluid processes and relatively permanent rules of belonging and association (like those formally imposed by the nation state). The tangible struggle to define its limits and range (sometimes even territories and borders), to create and sustain its rules and institutions through collective powers such as constitutional forms, political parties, the churches, the unions, neighborhood organizations, local governments, and the like, has proven central to the pursuit of alternatives to the selfishness of personalized market individualism. But, as many have recently pointed out, the re-making and re-imagining of 'community' will work in progressive directions only if it is connected *en route* to a more generalized radical insurgent politics. That means a radical project (however defined) must exist. The rule-making that ever constitutes community must be set against the rule-breaking that makes for revolutionary transformations.

The embeddedness and organized power community offers as a basis for political action is crucial even if its coherence requires democratically structured systems of authority, consensus, and 'rules of belonging'. Thus, although community 'in itself' has meaning as part of a broader politics, community 'for itself' almost invariably degenerates into regressive exclusions and fragmentations (what some would call negative heterotopias of spatial form). Means must therefore be found whereby we insurgent architects can reach out across space and time to shape a more integrated process of historical-geographical change beyond the limits typically defined by some community of common interest. The construction of collective identities, of communities of action, of rules of belonging, is a crucial moment in the translation of the personal and the political onto a broader terrain of human action. At the same time, the formation of such collectivities creates an environment and a space (sometimes, like the nation state, relatively stable and enduring) that shapes the political person as well as the ways in which the personal is and can be political.

4 Militant particularism and political action

The theory of 'militant particularism' argues that all broad-based political movements have their origins in particular struggles in particular places and times (see Harvey, 1996, Chapter 1). Many struggles are defensive – for example, struggles against plant closures or excessive exploitation of labor, the siting of noxious facilities (toxic waste dumps), the dismantling or lack of social or police protections, violence against women, the environmental transformations proposed by developers, the appropriation of indigenous resources by outsiders, attacks upon indigenous cultural forms, and the like. A widespread politics of resistance now exists, for example, to neoliberalism and capitalism throughout the world. But some forms of militant particularism are pro-active. Under capitalism this typically means struggles for specific group rights that are universally declared but only partially conferred (in the past this has usually meant the rights of entrepreneurs and owners of means of production to freely exercise their rights of ownership without restraint, but it has also extended to include the rights of slaves, labor, women, gays, the culturally different, animals and endangered species, the environment, and the like).

The critical problem for this vast array of struggles is to shift gears, transcend particularities, and arrive at some conception of a universal alternative to that social system which is the source of their difficulties. Capitalism (coupled with modernism and, perhaps, a Eurocentric 'Westernism') successfully did this *vis-à-vis* pre-existing modes of production, but the oppositional movements of socialism, communism, environmentalism, feminism, and even humanism and multiculturalism have all constructed some sort of universalistic politics out of militant particularist origins. It is important to understand how this universalization occurs, the problems that arise, and the role traditional utopianism plays.

Dialectics here is useful. It teaches that universality always exists *in relation to* particularity: neither can be separated from the other even though they are distinctive moments within our conceptual operations and practical engagements. The notion of justice, for example, acquires universality through a process of abstraction from particular instances and circumstances, but becomes particular again as it is actualized in the real world through social practices. But the orchestration of this process depends upon mediating institutions (those, for example, of language, law, and custom within given territories or among specific social groups). These mediating institutions 'translate' between particularities and universals and (like the Supreme Court) become guardians of universal principles and arbiters of their application. They also become power centers in their own right. This is, broadly, the structure set up under capitalism with the state and all of its institutions (now supplemented by a variety of international institutions such as the World Bank and the IMF, the United Nations, GATT and the World Trade Organization) being fundamental as 'executive committees' of capitalism's systemic interests. Capitalism is replete with mechanisms for converting from the particular (even personal) to the universal and back again in a dynamic and interactive mode. Historically, of course, the primary mediator has been the nation state and all of its institutions including those that manage the circulation of money.

No social order can, therefore, evade the question of universals. The contemporary 'radical' critique of universalism is sadly misplaced. It should focus instead on the specific institutions of power that translate between particularity and universality rather than attack universalism *per se*. Clearly, such institutions favor certain particularities (such as the rights of ownership of means of production) over others (such as the rights of the direct producers) and promote a specific kind of universal.

But there is another difficulty. The movement from particularity to universality entails a 'translation' from the concrete to the abstract. Since a violence attaches to abstraction, a tension always exists between particularity and universality in politics. This can be viewed either as a creative tension or, more often, as a destructive and immobilizing force in which inflexible mediating institutions (such as an authoritarian government apparatus) claim rights over individuals and communities in the name of some universal principle.

It is here that critical engagement with the static utopianism of spatial form (particularly its penchant for nostalgia) and the loosening of its hold by appeal to a utopianism of spatial-temporal transformation can keep open prospects for further change. The creative tension within the dialectic of particularity–universality cannot be repressed for long. Mediating institutions, no matter how necessary, cannot afford to ossify, and traditional utopianism is often powerfully suggestive as to institutional reforms. The dynamic utopian vision that emerges is one of sufficient stability of institutional and spatial forms to provide security and continuity, coupled with a dynamic negotiation between particularities and universals so as to force mediating institutions and spatial structures to be as open as possible.

At times, capitalism has worked in such a way (consider how, for example, the law gets reinterpreted to confront new socio-economic conditions and how the production of space has occurred throughout the long history of capitalism). Any radical alternative, if it is to succeed as it materializes, must follow capitalism's example in this regard. It must find ways to negotiate between the security conferred by fixed institutions and spatial forms on the one hand and the need to be open and flexible in relation to new socio-spatial possibilities on the other. Both Jefferson and Mao understood the need for some sort of 'permanent revolution' to lie at the heart of any progressive social order. The failure to acknowledge that imperative lies at the heart of the collapse of the Soviet Union and seriously threatens the United States. The perspective of a permanent revolution (in, for example, the production of spatial forms) must therefore be added to that of a long revolution as we reach for the principles of a spatiotemporal and dialectical utopianism.

5 Mediating institutions and built environments

The formation of institutions and built environments that can mediate the dialectic between particularity and universality is of crucial importance. Such institutions typically become centers for the formation of dominant discourses as well as centers for the exercise of power. Many of them – medical care, education, financial affairs, and the state – cultivate a special expertise in the same way that built environments of different sorts facilitate possibilities for social action in some directions while limiting others. Many institutions (e.g. local governments and the state) are organized territorially and define and regulate activity at a particular spatial scale. They can translate militant particularism into an institutionalized spatial order designed to facilitate or repress certain kinds of social action and thereby influence the ways in which the personal can be political, encouraging some (like entrepreneurial endeavor, say) and discouraging others (like socialist communes).

Much the same can be said of the built environments that get constructed. Consider, for example, the form and style of urbanization and the consequences that flow therefrom. How can the personal be openly political when environmental conditions inhibit the free exploration of radically different lifestyles (such as living without an automobile or private property in Los Angeles)? The uneven conditions of geographical development that now prevail in Baltimore do not allow the personal to be political in anything other than rather restrictive ways (equally repressive, though in every different ways, for the affluent child of suburbia as for the child of inner-city poverty).

The creation of mediating institutions is deeply fraught and frequently contested (as one might properly expect). The chief difficulty is to bring multiple militant particularisms (in the contemporary US this might mean the aspirations of radical ecologists, the chamber of commerce, ethnic or religious groups, feminists, developers, class organizations, bankers, and the like) into some kind of institutional relation to each other without resort to arbitrary authority and power. The Porto Alegre experience (see Abers, 1998) suggests that this sort of thing can be done. But decisions have to be made and arbitrary authority and power are invariably implicated in the process. With the best will in the world these cannot be eliminated. The effect is to render the mediating institutions sites of power and thereby sources of distinctive discourses and constructions which can be organized in a system of dominance which individual persons find hard to resist let alone transcend. The capture or destruction of mediating institutions (such as the state, the financial sector, education) and the re-shaping of built environments has often, therefore, been the be-all and end-all of insurgent radicalism. While this is one crucial theater in the long frontier of insurgent politics, it is far from being the whole of the story.

6 Translations and aspirations

The insurgent architect with a lust for transformative action must be able to translate political aspirations across the incredible variety and heterogeneity of socio-ecological and political-economic conditions. He or she must also be able to relate different discursive constructions and representations of the world (such as the extraordinary variety of ways in which environmental issues are discussed). He or she must confront the conditions of and prospects for uneven geographical developments. The skills of translation here become crucial. For James Boyd White (1990, 257–64):

> [Translation means] confronting unbridgeable discontinuities between texts, between languages, and between people. As such it has an ethical as well as an intellectual dimension. It recognises the other – the composer of the original text – as a center of meaning apart from oneself. It requires one to discover both the value of the other's language and the limits of one's own. Good translation thus proceeds not by the motives of dominance and acquisition, but by respect. It is a word for a set of practices by which we learn to live with difference, with the fluidity of culture and with the instability of the self.   (257)

> We should not feel that respect for the other obliges us to erase ourselves, or our culture, as if all value lay out there and none here. As the traditions of the

other are entitled to respect, despite their oddness to us, and sometimes despite their inhumanities, so too our own tradition is entitled to respect as well. Our task is to be distinctively ourselves in a world of others: to create a frame that includes both self and other, neither dominant, in an image of fundamental equality. This is true of us as individuals in our relations with others, and true of us as a culture too, as we face the diversity of our world . . . This is not the kind of relativism that asserts that nothing can be known, but is itself a way of knowing: a way of seeing one thing in terms of another. Similarly it does not assert that no judgments can be reached, but is itself a way of judging, and of doing so out of a sense of our position in a shifting world.   (264)

This, in itself, has its own utopian ring. It is not hard to problematize such an argument, as Said did so brilliantly in *Orientalism*, as the power of the translator (usually white male and bourgeois) to represent 'the other' in a manner that dominated subjects (orientals, blacks, women, etc.) are forced to internalize and accept. But that historical understanding itself provides a hedge against the kinds of representational repressions that Said and many feminists have recorded. This links us back to how the personal is always political. As White notes: 'to attempt to "translate" is to experience a failure at once radical and felicitous: radical, for it throws into question our sense of ourselves, our languages, of others; felicitous, for it releases us momentarily from the prison of our own ways of thinking and being' (1990, 257). The act of translation offers a moment of liberatory as well as repressive possibility. The architects of spatiotemporal utopianism must be open to such possibilities.

But as real architects of our future we cannot engage in endless problematization and never-ending conversations. Firm recommendations must be advanced and decisions taken, in the clear knowledge of all the limitations and potentiality for unintended consequences (both good and bad). We need to move step by step towards more common understandings. And this for two compelling reasons. First, as Zeldin (1994, 16) among others remarks, we know a great deal about what divides people but nowhere near enough about what we have in common. The insurgent architect has a role to play in defining commonalities as well as in registering differences. But the second compelling reason is this: without translation, collective forms of action become impossible. All potential for an alternative politics disappears. The fluid ability of capitalists and their agents to translate among themselves using the basic languages of money, commodity, and property (backed, where necessary, with the theoretical language of a reductionist economics) is one of their towering class strengths. Any insurgent oppositional movement must do this just as well if not better. Struggle as we may, it is impossible to conduct politics without an adequate practice of translation. If reductionism of the Wilsonian sort is rejected, then the only option is translation. Thomas Kuhn, in his *Structure of Scientific Revolutions*, considers translation (rather than reduction) as the privileged and perhaps sole means by which fundamentally different paradigms of scientific knowledge might be related, and Judith Butler (1998, 38), under pressure from her critics as to the fragmenting effect of identity politics, argues:

> Whatever universal becomes possible – and it may be that universals only become possible for a time, 'flashing up' in Benjamin's sense – will be the result of the difficult labour of translation in which social movements offer up their points of convergence against a background of ongoing social contestation.

The omnipresent danger in any dialectical utopianism is that some all-powerful center or some elite comes to dominate. The center cajoles, bullies, and persuades its periphery into certain modes of thought and action (much as the United States has done since World War II, culminating in the infamous Washington consensus through which the United States sought to institutionalize its hegemonic position in the world order by gaining adhesion of everyone to certain universal principles of political-economic life). As opposed to this, the democratic and egalitarian rules of translation should be clear. But so should the universal principles that, however much they merely 'flash up' as epiphenomena, emerge from the rich experience of translation to define what it is we might have in common.

7 The moment of universality: On personal commitments and political projects

The moment of universality is not a final moment of revelation or of absolute truth. I construe it, in the first instance, as a moment of existential decision, a moment of 'either/or' praxis, when certain principles are materialized through action in the world. It is, as it were, a nature-imposed condition of our species existence that we have to make decisions (individually and collectively) and we have to act upon them. The moment of universality is the moment of choosing, no matter how much we may reserve judgement on our actions afterwards. How we come to represent those decisions to ourselves in terms of principles or codes of action that act as guidelines for future decisions is an important cultural value that gains power over us as it becomes instantiated in discourses and institutions. It is here that abstract universal principles operate as plays of power.

Universals cannot and do not exist, however, outside of the political persons who hold to them and act upon them. They are not free standing nor do they function as abstracted absolutes that can be brought to bear upon human affairs for all times and places. They are omnipresent in all practices. But to the degree that we begin to shape and order them for given purposes they take on the guise of abstract principles (even written codes and laws) to which we adhere. And if we find in them successful guides to action (as we do, for example, within the corpus of scientific understandings) so they shape our world view and become institutionalized as mediating discourses. They tend to cluster and converge as dominant paradigms, as hegemonic discourses, or as pervasive ethical, moral, or political-economic principles that inform our beliefs and actions. They become codified into languages, laws, institutions, and constitutions. Universals are socially constructed not given.

While social construction can betoken contestation, it is more often the case that the dominant principles handed down to us so limit our conceptions as to inhibit alternative visions of how the world might be. A wide range of possible universal and unifying principles has in fact been bequeathed to us (the fruit of long and often bitter experience). But, as many commentators point out (usually with critical intent), many of these principles have their origins in the Western Enlightenment when theorists of the natural and social order had, unlike now, no hesitation in expressing their opinions as universal truths and propositions. It is fashionable in these times to denigrate these (at least in the humanities) while at the same time leaving crude versions of them fully in play in society in general. But we can never do without universals of some sort. We can, of course, *pretend* to do without them. Much of what now passes for radical argument in the humanities and some segments of the social sciences resorts to much dissimulation and opacity (when it is not engaged in downright chicanery) with respect to this point.

It is therefore important at the outset to exhume the traces of universal principles expressed in the ways the personal is and can be political. This is so because without certain criteria of judgement (explicit or implied), it is impossible to distinguish between right and wrong or between progressive and regressive lines of political action. The existential moment of do I or do I not support this or that line of action entails such a judgement. Even though I may prefer not to make it, not to decide is in itself a form of decision (one that many Americans now prefer at the ballot box with specific consequences). So though the moment of universality is not the moment of revelation, it is the moment of *judgement and decision* and these willy-nilly entail *expression* of some universal whether we like it or not. It is only in these terms that we are able to say that *this* form of insurgent politics (embedded, say, in a movement for environmental justice) is progressive and worthy of support while *that* form of insurgent politics (like the militia movement in the Michigan woods) is not. The moment of the universal is, therefore, the moment of political judgement, commitment, and material praxis.

For this reason it is, paradoxically, the moment that gets argued over in the most abstract of terms. In effect, we seek to create a generalized discourse about rights and wrongs, about moral imperatives and proper and improper means and ends, through which we try to persuade ourselves as well as others to certain consistent lines of action, knowing full well that each of us is different and that no particularity is exactly the same as any other.

Such arguments can easily seem redundant, but when connected back across all the other theaters of action on the long frontier of insurgent politics, they can acquire a stronger force and even provide some sort of political and emotive thread that helps us recognize in what ways the personal, the collective, the mediating institutions can relate to each other in dynamic ways through the activities of the translator and the insurgent architectural imagination. Furthermore, it is also the case that universals draw their power and meaning from a conception of species being (it is only in terms of species rights that universal principles of conduct can make sense). It then also follows (as I argued in Chapter 11) that acceptance of some sort of 'unity of science' is a necessary condition for the promulgation of universality claims. Conversely, discussions of universality crucially depend upon critical engagement with notions of species being and the unity of science.

So what universals might we currently embrace as meaningful ideals upon which to let our imaginations roam as we go to work as insurgent architects of our future? I have already referred in Chapter 5 to the United Nations Declaration of Human Rights as a document that expresses such universal principles in problematic but to some degree persuasive terms. The application of these principles has often been contested and their interpretation has had to be fought over in almost every particular case. Can we add or re-formulate those universals in interesting ways? My own preferred short-list of universal rights worthy of attention runs as follows:

1. The right to life chances
   This entails a basic right to sustenance and to elemental economic securities. Food security would be the most basic manifestation of such rights, but a general system of entitlements – as Sen (1982) would call them – is also fundamental. This re-affirms the UN Declaration (Article 23, Section 3) that 'everyone has the right to just and favourable remuneration ensuring himself and his family an existence worthy of

human dignity, and supplemented, if necessary, by other means of social protection.' The universal right to a 'living wage' and to adequate social security is one way to both demand and problematize such a universal package of rights.

2. The right to political association and 'good' governance
    Individuals must have the right to associate in order to shape and control political institutions and cultural forms at a variety of scales (cf. Articles 20 and 21 of the UN Declaration). The presumption is that some adequate definition can be found for properly democratic procedures of association and that collective forms of action must offer reasonable protections to minority opinions. The presumption also exists that some definition of 'good' governance can be found, from the local to the global level. Here, too, the demand highlights problems and differences (the definition of 'good governance' is far from homogeneous) at the same time as it takes up universalizing claims. But individuals plainly should have rights to produce their own spaces of community and inscribe their own rules therein, even as limitations on such rights become critical to restrict the narrow exclusions and the internal repressions to which communitarianism always tends.

3. The rights of the direct laborers in the process of production
    The rights of those who labor to exercise some level of individual and collective control over labor processes (over what is produced as well as over how it shall be produced) is crucial to any conception of democracy and freedom. Long-standing concerns over the conditions of labor and the right of redress in the event of unreasonable burdens or sufferings (such as those that result in shortened life expectancy) need to be reinforced on a more global scale. This entails a demand for the radical empowerment of the laborer in relation to the production system in general (no matter whether it is capitalist, communist, socialist, anarchist, or whatever). It also highlights respect for the dignity of labor and of the laborer within the global system of production, exchange, and consumption (on this point, at least, a variety of Papal Encyclicals as well as the UN Declaration provide supportive materials).

4. The right to the inviolability and integrity of the human body
    The UN Declaration (Articles 1 to 10) insists on the right to the dignity and integrity of the body and the political person. This presumes rights to be free from the tortures, incarcerations, killings, and other physical coercions that have so often been deployed in the past to accomplish narrow political objectives. The right of women to control their own reproductive functions and to live free of coercions and violence (domestic, cultural, and institutionalized) must also lie at the core of this conception. Violence against women and the subservience of women to patriarchal and paternalistic systems of domination has become a major issue for which universal rights claims have become deeply plausible and

compelling (though often in conflict with claims for autonomy of cultural traditions).

5. Immunity/destabilization rights

   Everyone has the right to freedom of thought, conscience, and religion, according to the UN Declaration (Articles 18 and 19). On this point the Declaration is definitive and clear. But I here think Unger's (1987b, 524–34) argument for a system of immunity rights that connects to a citizen's rights to destabilize that which exists is even stronger, for it insists on the right to critical commentary and dispute without fear of retaliation or other loss. It is only through the exercise of such rights that society can be both re-imagined and re-made (Unger's arguments on this point are persuasive).

6. The right to a decent and healthy living environment

   From time to time legislation in particular countries has been predicated on the right of everyone to live in a decent and healthy living environment, one that is reasonably free from threats and dangers and from unnecessary hazards (particularly those produced through human activities, such as toxic wastes, dirty air, and polluted waters). The spreading cancers of environmental injustice throughout the world and the innumerable consequences for human health and well-being that flow from environmental degradations (both physical and social) indicate a terrain where the proper establishment of universal rights is imperative, even if it is surely evident that the meaning, interpretation, and application of such rights will be difficult to achieve.

7. The right to collective control of common property resources

   The system of property rights by which capitalism has typically asserted its universalizing claims (actively supported in Article 17 of the UN Declaration) is widely understood as both defective and in some instances destructive with respect to our physical and social world. This is nowhere more apparent than in instances of common property resources (everything from genetic materials in tropical rain forests to air, water, and other environmental qualities including, incidentally, the rights to control built environments for historical, cultural, or aesthetic reasons). The definition of such resources and the determination of who is the 'collective' in whose name rights of control will be vested are all deeply controversial issues. But there are widespread arguments now for alternative systems of property rights to those implied in a narrowly self-serving and myopic structure of private property rights that fail to acknowledge any other form of public or collective interest to that given through a pervasive market (and corporation-dominated) individualism.

8. The rights of those yet to be born

   Future generations have a claim upon us, preferably to live in a world of open possibilities rather than of foreclosed options. The whole rhetoric of sustainable environmental development rests on some sense (however vague and undefined) of responsibilities and obligations that stretch beyond the ken of our own immediate interests. *In extremis*, this right also recognizes our volitional role in the evolutionary process and our responsibilities not only to our own species but also to the innumerable others whose prospects for survival depend upon our actions (see Item 11).

9. The right to the production of space

The ability of individuals and collectivities to 'vote with their feet' and perpetually seek the fulfillments of their needs and desires elsewhere is probably the most radical of all proposals. Yet without it there is nothing to stop the relative incarceration of captive populations within particular territories. If, for example, labor had the same right of mobility as capital, if political persecution could be resisted (as the affluent and privileged have proven) by geographical movement, and if individuals and collectivities had the right to change their locations at will, then the kind of world we live in would change dramatically (this principle is stated in Article 14 of the UN Declaration). But the production of space means more than merely the ability to circulate within a pre-ordained spatially structured world. It also means the right to reconstruct spatial relations (territorial forms, communicative capacities, and rules) in ways that turn space from an absolute framework of action into a more malleable relative and relational aspect of social life.

10. The right to difference including that of uneven geographical development

The UN Declaration (Articles 22 and 27) states that everyone should be accorded 'the economic, social and cultural rights indispensable for his dignity and the free development of his personality' while also pointing to the importance of the right 'freely to participate in the cultural life of the community' and to receive protection of 'the moral and material interests resulting from scientific, literary or artistic production.' This implies the right to be different, to explore differences in the realms of culture, sexuality, religious beliefs, and the like. But it also implies the right for different group or collective explorations of such differences and, as a consequence, the right to pursue development on some territorial and collective basis that departs from established norms. Uneven geographical development should also be thought of as a right rather than as a capitalistically imposed necessity that diminishes life chances in one place in order to enhance them elsewhere. Again, the application of such a principle in such a way that it does not infringe upon others in negative ways will have to be fought over, but the statement of such a principle, like that of the living wage, provides a clear basis for argument. The recent UN extension of cultural rights (particularly those specified in Article 27 of the original UN Declaration) to encompass those of minorities (cf. Phillips and Rosas, 1995) provides an initial opening in this direction.

11. Our rights as species beings

This is, perhaps, the vaguest and least easily specifiable of all rights. Yet it is perhaps the most important of them all. It must become central to debate. If we review our position in the long history of biological and social evolution, then plainly we have been and continue to be powerful evolutionary agents. If we are now entering a phase of volitional and conscious interventions in evolutionary processes (interventions that carry with them enormous risks and dangers), then we must necessarily construe certain universals to both promote and regulate the way we might engage upon such interventions. We all should have the right freely to explore the relation to nature and the transformative possibilities inherent in our species being in creative ways. This means the right to explore the possibility of different combinations of our evolutionary repertoire – the powers of cooperation, diversification, competition, the production of nature and of different dimensionalities to space and time. But that right to free experimentation (made much of by Unger) must also be tempered by duties, responsibilities, and obligations to others, both human and non-human, and it most certainly must accord strong protections against the potential powers of a non-democratic elite (or a capitalist class) to push us down technological, social, and evolutionary pathways that represent narrow class interests rather than human interests in general. Any concept of 'species interests' will inevitably be riven by rampant divisions of class, gender, religion, culture, and geography. But without some sense of where our common interests as a species might lie, it becomes impossible to construct any 'family of meanings' to connect or ground the incredible variety of partial claims and demands that make our social world such an interestingly divided place. On this point Naess and Rothenberg (1989, 164–70) have much to offer, by insisting that 'the universal right to self-unfolding' is related to the recognition of that same right across all species, and that 'the unfolding of life' in general is as important as the unfolding of our own personal trajectories of self-discovery and development.

This interlocking and oftentimes conflictual system of universal rights, I insist, is not the be-all and end-all of struggle, but a formative moment in a much more complicated social process directed towards socio-ecological change that embraces all the other distinctive theaters of social action. But the insurgent architect has to be an advocate of such rights. At the same time he or she must clearly recognize that their formulation arises out of social life and that they remain otiose and meaningless unless brought to bear in tangible ways upon mediating institutions, processes of community formation, and upon the ways in which the personal is construed and acted upon as the political.

8 Shaping socio-ecological orders

The dialectical utopianism to which I aspire requires the perspective of a long and permanent historical-geographical revolution. Thinking about transformative political practices as manifestations of a dialectical and spatiotemporal utopianism is helpful. But it will only be so if we understand how activity and thought in the different theaters of social action relate, combine, and dissolve into each other to create an evolving totality of social action.

Unfortunately, much that passes for imaginative architectural and political practice often stays immobilized in only one or two of the theaters I have here defined. Our mental and practical divisions of labor and of perspectives are now so deeply ingrained in everything we do that it becomes impossible for any one of us to be fully present in much more than one of these theaters of thought and action at any one time. The problem is not that this cannot work. Indeed, it may work far too well as it so patently has in the past (as, for example, dominant mediating institutions use divisions of mental and practical labors to dictate the terms of universality and the ways it is admissible for the personal to be political). The errors of that past always threaten to return and haunt us (though perhaps in different ways). By seeing the seven moments I have described as integral to each other, by recognizing how they are all internally related, and by seeking to flow our analysis, our thinking, and our practices across their entire range, we may better situate our capacities as insurgent architects of some alternative possible dynamics. Any aspiring insurgent architect must learn, in association with others, to collate and combine action on all fronts. Universality without the personal is abstract dogma if not active political hypocrisy. Community without either the personal or universal becomes exclusionary and fascistic. Mediating institutions that consolidate their powers and oppress the personal or translate universals into bureaucratic systems of despotism and control subvert the revolutionary impulse into state authoritarianism. The translator who assumes omnipotence represses. The great individual (the architect/philosopher) who becomes detached from the masses and from daily life becomes either an irrelevant joke or an oppressive and domineering figure on the local if not on the world stage.

It is open dialogue and practical interactions across theaters on this long frontier that counts. And it is to dialectics rather than Wilsonian reductionism that we must appeal to make the connexions, however putatively, across these different scales. Only then can the impulse towards dialectical utopianism be prevented from dissolving into the arid and ultimately self-destructive utopianism of either closed spatial form or of temporal processes of perpetual creative destruction.

But aspirations must be tempered by a sense of limitations and of vulnerability. There are necessary limits to even the most vaunting of ambitions. If, as I have argued, dialectical utopianism must be effectively grounded in historical-geographical realities and achievements, if, to return to Marx's celebrated formulation, we can always aspire to make our own historical geography but never under historical and geographical conditions of our own choosing, then the leap from the present into some future is always constrained, no matter how hard we struggle to liberate ourselves from the three basic constraints of (1) where we can see it from, (2) how far can we see, and (3) where we can learn it from.

And as we make that leap, we also have to acknowledge that it is a speculative leap into the unknown and into the unknowable. There is a level at which, no matter how hard we try, we simply cannot know with certainty what kind of outcomes will emerge. Both the social and the ecological orders, particularly when taken together, are open and hetero-genous to the point where their totality can never quite be grasped let alone manipulated into predictable or stable states. No matter how hard we try to construct and reconstruct the socio-ecological order to a given plan, we inevitably fall victim not only to the unexpected consequences of our own actions but also to evolutionary contingencies (those 'accidents' to which Marx referred) that impinge upon us at every twist and turn and at every scale. It is precisely for this reason that the ideals of community, of utopias of spatial form, exercise such an attraction because they depict a closed world of known certainties and rules where chance and contin-gency, uncertainty and risk, are resolutely locked out.

Herein lies perhaps the most difficult of all barriers for the insurgent architect to surmount. In facing up to a world of uncertainty and risk, the possibility of being quite undone by the consequences of our own actions weighs heavily upon us, often making us prefer 'those ills we have than flying to others that we know not of.' But Hamlet, beset by angst and doubt and unable to act, brought disaster upon himself and upon his land by the mere fact of his inaction. It is on this point that we need to mark well the lessons of capitalist historical geography. For that historical geography was created through innumerable forms of speculative action, by a preparedness to take risks and be undone by them. While we laborers (and philosophical underlaborers) may for good reasons 'lack the courage of our minds,' the capitalists have rarely lacked the courage of theirs. And, arguably, when they have given in to doubt they have lost their capacity to make and re-make the world. Marx and Keynes, both, understood that it was the 'animal spirits,' the speculative passions and expectations of the capitalist (like those that Zola so dramatically depicted) that bore the system along, taking it in new directions and into new spaces (both literal and metaphorical). And it is perhaps no accident that architecture as a supremely speculative and heroic profession (rather than as either a Platonic metaphor or a craft) emerged in Italy along with the merchant capitalists who began upon their globalizing ventures through commercial speculations in the fifteenth and sixteenth centuries. It was that speculative spirit that opened up new spaces for human thought and action in all manner of ways.

The lesson is clear: until we insurgent architects know the courage of our minds and are prepared to take an equally speculative plunge into some unknown, we too will continue to be the objects of historical geography (like worker bees) rather than active subjects, consciously pushing human possibilities to their limits. What Marx called 'the real movement' that will abolish 'the existing state of things' is always there for the making and for the taking. That is what gaining the courage of our minds is all about.

Appendix
Edilia, or
'Make of it what you will'

Sometime in 1888, Ebenezer Howard read Edward Bellamy's just-published utopian novel *Looking Backward*. He did so at one sitting and was 'fairly carried away' by it. The next morning he:

> went into some of the crowded parts of London, and as I passed through the narrow dark streets, saw the wretched dwellings in which the majority of the people lived, observed on every hand the manifestations of a self-seeking order of society and reflected on the absolute unsoundness of our economic system, there came to me an overpowering sense of the temporary nature of all I saw, and of its entire unsuitability for the working life of the new order – the order of justice, unity and friendliness.

Howard fused the two sentiments. He sought a way to realize Bellamy's vision and promote 'the order of justice, unity and friendliness' that he found so lacking in the London of his day. In 1898, he published, at his own expense (publishers and magazine editors having proven indifferent or hostile), what was later to be called *Garden Cities of Tomorrow*. And so the 'new towns' movement was born, a movement that turned out to be one of the most important interventions in urban re-engineering in the twentieth century.

The end of another century. I walk the streets of Baltimore and am even more appalled than Howard at the lack of justice, unity and friendliness. I say 'more appalled' because now the inequalities are so striking, so blatantly unnecessary, so against any kind of reason, and so accepted as part of some immutable 'natural order of things,' that I can scarcely contain my outrage and frustration. The immense pool of talents of a whole generation has been drained away into festering pits of alienation and anomie, anger and despair, blasé disinterest.

Is there no alternative? Where is that inspiring vision of the sort that Bellamy provided? It is, alas, all too fashionable in these times to proclaim the death of Utopia, to insist that utopianism of any sort will necessarily and inevitably culminate in totalitarianism and disaster. Naturally enough, our urban problems, when seen through the prism of such cynicism, seem intractable, immune to any remedy within the grasp of us mere mortals. There is, we conclude, 'no alternative' on this earth. We either look for remedies in the after-life or, Hamlet-like, prefer meekly to accept those urban ills we have 'than fly to others that we know not of.'

I retire to my study and browse among my books. I read the critical legal scholar Roberto Unger who complains that all of us have become 'helpless puppets of the institutional and imaginative worlds we inhabit.' We seem unable to think outside of existing structures and norms. We are torn, he says, 'between dreams that seem unrealizable' (the fantasy worlds given to us by the media) 'and prospects that hardly seem to matter' (daily life on the street). I put Unger's book down and pick up another by the philosopher Ernst Bloch who wonders why it is that 'possibility has had such a bad press.' There is, he sternly warns, 'a very clear interest that has prevented the world from being changed into the possible.' So there it is. 'There is no alternative.' I hark back to how often and to what political effect Margaret Thatcher repeatedly used that phrase. I fall into a reverie. 'No alternative, no alternative, no alternative' echoes in my mind. It pummels me into sleep where a whole host of utopian figures return to haunt me in a restless dream. This is what they tell me.

. . .

It will doubtless surprise you to know that by 2020 the revolution was over. In just seven years society underwent such a radical restructuring that it became unrecognizable.

The collapse began early in 2013. It centered on the stock market which rose in the first decade of the century to 85,000 points on the Dow – a level necessary, as many analysts pointed out, to properly service the aspirations of baby boomers like you whose accumulated savings were fueling the rise.

But there were many other signs of trouble. Global warming had kicked in with a vengeance by 2005, creating environmental havoc and crop failures in certain regions. This unleashed a pandemic of infectious diseases, created millions of environmental refugees, and generated a raft of increasingly burdensome insurance claims. Social inequality was severe enough in your time (remember how in 1990, 358 billionaires commanded assets equivalent to 2.7 billion of the world's poorest people?). But by 2010, ten percent of the world's population controlled 98 percent of its income and wealth.

Much of this wealth was spent on building formidable barriers against the poor (far stricter than your modest gated communities). Building barriers to keep themselves out was, indeed, a primary occupation for those who were lucky enough to find employment as common laborers in those years. But the higher the barriers, the more the wretched of the earth seemed to constitute an ever-growing danger.

But it was the stock market crash that unhinged the world. Nobody knows quite how it happened (does one ever?). Stock markets in Russia suddenly collapsed but by then the world was used to events of this sort (recall Indonesia or even Russia in 1998?). The general expectation was that another round of forced austerity (pushed by the world's central banks) coupled with financial jiggery-pokery would be sufficient to cure the problem.

However, some of the affluent baby-boomers decided it was time to cash in. As they did so the market fell and the more it fell the more others tried to cash in before it was too late.

For some four months, governments and central banks pumped in enough liquidity to hold markets steady (the Dow held at around 50,000). But then the world was awash with useless paper currencies. Inflation accelerated so that cans of tuna and bags of rice became more legitimate forms of currency than dollars, yen or euros. Interest rates soared into the thousand percent range.

Firms – even profitable ones – went bankrupt in the financial meltdown; and unemployment (in those places where such measures still mattered) soared to levels well beyond those ever before experienced touching even the affluent (you professors really came off badly).

Political power slowly dissolved under the crushing weight of rapid currency depreciations. Governments fell into disrepute and disarray. If the best government is limited to that which money can buy, then, it stands to reason, worthless money buys worthless government.

Private property rights and the artificial scarcities they underpin began to erode before the sheer force of human want and need. The law lost much of its meaning as the power to enforce private contracts evaporated in a maelstrom of entangling indebtedness. The legal system slowly surrendered to brute force since police power could not discriminate between the rights and wrongs of so many unenforceable contracts.

The dam finally broke in early fall of 2013. By year's end the Dow stood at less than 2,000. Pension funds and insurance companies went under as did banks and most other financial institutions. Stock markets disappeared. Paper wealth meant nothing. Baby boomers like you lost their financial security (your pension rights completely disappeared for example). Everyone lost because in the great reform of 2005 people had foolishly accepted that privately funded social security schemes were better than those of governments. This had fueled the march of the stock market just as it was now a victim of its collapse.

The military takeover of 2014 was violent. A tough hierarchy of command and administration was devised. Military law and order was ruthlessly imposed upon the world. Dissidents were rounded up and the wretched of the earth were pushed back into their corners and left to suffer and die under the fearful and malevolent eye of the military apparatus.

The generals issued proclamations stating that the central problem was overpopulation. Too many people were chasing too few resources. Carrying capacities were everywhere being violated and optimal populations (calculated by certain ecologists of your time to stand at no more than one hundred million for the United States) were being grossly exceeded. Regrettable as it may seem, they said, the checks to population growth that Malthus had described in his famous essay on population (published first in 1798), principally those of famine, disease, and war, provided the only proper means to bring population and resources back into balance. Only when nature ('red in tooth and claw') had done its work could sustainability be achieved and the natural order be restored.

Thus was the coming anarchy predicted, justified, and managed.

The military did, however, lay the groundwork for a complete reorganization and rationalization of social life.

By 2010 the data banks on individuals compiled by credit institutions (already far more highly developed in your times than even you suspected) had been consolidated into one mass surveillance system. To protect themselves, the rich had insisted upon the implantation of electronic surveillance devices into the bodies of those that served them to ensure that nobody unwanted ever ventured into their presence. It became possible to scan individuals and obtain an instant 'bio' on each (imagine the uses to which this was put!).

The military universalized this system. Everyone they could round up was electronically implanted and coded. Everyone's location could be monitored from space. This was George Orwell's 'Big Brother' raised to the nth power. Among the revolutionary minded it was regarded as something to be smashed in the name of personal liberty and freedom.

The military also took every technology that lay to hand plus a few more to create a remarkable communications system immune to attack from even the most talented hackers. They similarly established highly efficient and far more ecologically sound systems of transportation. Used for ever tighter forms of surveillance and control, these systems could later be converted to satisfy different needs.

Lacking any clear moral authority or popular legitimacy, the military entered into alliances with religious powers, creating a global system of governance that amounted to militarized theocracies, divided regionally according to religious affiliations (a division that initially sparked massive and disruptive population movements that made those of 1947 on the Indian sub-continent seem miniscule).

These military theocracies even negotiated, briefly, a period of world unitary governance, calling back an aging George Soros to be the first president of the Concert of the World. But that soon fell apart. The more centralized and hierarchically tight each military theocracy became, the more Darwinian the struggle between them, sparking costly wars of attrition all around the world.

At first, such struggles, invariably depicted on both sides as a struggle against some other evil empire, helped consolidate internal controls by mobilizing nationalistic fervors and religious hatreds. But the falling apart of internal powers of provision made it harder and harder for central authorities to exercise control within localities where people waged the bitter but crucial economic struggle for daily survival.

As the authorities appropriated more and more resources, so local discontents burgeoned. Mini-movements of opposition sprang up everywhere, militating against the corruptions and policies of hierarchical powers, criticizing the authorities for failing to deal with pressing problems of survival. These movements began to organize themselves on a localized basis. They built pioneering collective structures for both survival and resistance. But, unarmed, they were often subject to violent repression.

How exactly it happened remains obscure, but in 2019 these disparate and fragmented movements suddenly came together (some later said through the machinations of some secret society of revolutionary organizers, though there was little evidence for that at the time).

The wretched of the earth spontaneously and collectively rose up. They created a massive movement of non-violent resistance, silently occupying more and more spaces of the global economy, while issuing rapid-fire demands for greater equality, the disbanding of military power, and the impeachment of military and religious leaders.

The authorities viewed this as collective madness. Frustrated and frightened, they launched into irrational and uncontrollable violence, lashing out in fear and loathing at their own peoples as well as against each other. Such actions confirmed rather than diminished the determination of the dispossessed to take control. And in many of those enclaves of opposition and self-support, peoples had already achieved levels of conviction, understanding, and solidarity unparalleled in human history.

Sickened by the egregious violence, elements within the military deserted, as did many holy men who had long sympathized with the people against the authorities. Hierarchical religion and militarism began to collapse. Their corrupt reliance upon each other and their blatant hypocrisy and venality proved their undoing.

Many scientists, doctors, and technicians dropped their support of the militarized theocracies and put their skills at the service of the new movement, at first subversively but then openly as liberated territories began to be defined. Deserting their positions as privileged hacks of a corrupt military theocracy, they launched a movement to reassert the emancipatory and humanizing mission of science, medicine, and learning.

This proved an extraordinary cultural moment (and one which was to be memorable for everyone who participated). While religious authority self-destructed and the hierarchical powers of an absolutist militarized science fell apart, the two most powerful forces with which humans are endowed, those of spiritual commitments and scientific enquiry, merged to found a humanized and politically aware scientific sensibility that was to be the cornerstone of political revolution.

Many intellectuals and artists joined the uprising. Some managers and technicians (particularly those in charge of idle factories) became enamored of the prospects of real change and sporadically led the occupation of their workspaces (factories, farms, and offices), committing themselves to put the productive apparatus back to work for different social ends.

After the military theocracies had been reduced to an illegitimate rump of power, brandishing weapons of mass destruction – and on a couple of terrible occasions actually using them – a peaceful, non-violent mass movement, led almost entirely by women, swept across the globe.

This movement disarmed both the military and the motley array of marauding bands, mafiosi, and vigilante groups that had here and there sprung up to fill the vacuum left by waning military power. Such marauding bands threatened to form an entirely new anarcho-nihilistic social order in which male violence and patriarchy would be the primary source of political authority.

The women's movement that counteracted this threat arose among the billion or so women who, by 2010, formed the majority of the proletariat. It was this feminized proletariat that was to be the agent of historical transformation. These women worked under insufferable conditions of oppression and continued to be lumbered with all the key responsibilities of reproduction (while being excluded, particularly under the military theocracies, from public power).

They everywhere came to call themselves 'The Mothers of Those Yet to be Born' (a name taken from the first manifesto of such a movement in Buenos Aires published in June 2019). They went from house to house and place to place, destroying every weapon and firearm they could find, ultimately galvanizing a whole army of newly recruited technicians – both men and women – to neutralize and eliminate all weapons of violence and mass destruction. It was a non-violent movement, combining passive resistance and mass action. It was the most powerful blow of all struck for social equality and mutual respect, albeit in the midst of revolutionary turbulence.

This movement proved the catalyst that turned the world away from centralized hierarchies of power to a politics of egalitarian mass force connecting localities, individuals, and all manner of social groups into an intricate and interactive network of global exchange.

By 2020 much of the world was disarmed. Military and religious authority had slowly asphyxiated in a deadly embrace. All those interests that would prevent the realization of the possible were subdued. People could think about, discuss, and communicate their alternative visions.

And this is the society that the wretched of the earth led by *The Mothers of Those Yet to Be Born* dreamed up and actually made, in alliance with the scientists, intellectuals, spiritual thinkers, and artists who had liberated themselves from their deadening political and ideological subservience to class power and to military-theocratic authority.

. . .

The basic unit of habitation is called a *hearth*. It is comprised of anywhere between twenty and thirty adults and whatever children are attached to parenting collectives called *pradashas* (on which more anon) within it. Each *hearth* forms a collective living arrangement organized as a common economy for mutual self-support.

The members of the *hearth* eat and work together, arriving at collective decisions as to how to organize themselves internally and how to 'make a living' through exchanges with other *hearths*. A *neighborhood* comprises some ten or so *hearths* and a larger organizational unit, called an *edilia*, loosely coordinates activities across two hundred or more *hearths* (roughly sixty thousand people). The largest continuous political unit is a *regiona* comprising anywhere from twenty to fifty *edilias* (at most three million people). The aim is for this to form a bioregion of human habitation that strives to be as self-sufficient as practicable, paying close attention to environmental problems and sustainabilities.

Beyond this there is the *nationa*, which is a loosely organized federation of *regionas* collected together for purposes of mutual barter and trade. It typically comprises at least a couple of *regionas* in each of the tropical, subtropical, temperate, and sub-arctic parts of the world respectively with a similar diversification between continental and maritime, arid and well-watered *regionas*. The statutes of federation are periodically renegotiated and *regionas* sometimes shift from one *nationa* to another as they see fit. Furthermore, new *nationas* can form at will while others dissolve so there is no fixed scale of population nor even any fixed political organization beyond the statutes of federation.

Since the free flow of goods across these ecologically defined *regionas* is fundamental to the support of a reasonable standard of living (guaranteeing as far as possible against localized famines and shortages) it also means an equally free flow of people. As a consequence, the degrees of racial, ethnic, and cultural mixing (already much jumbled up during the revolutionary period) make any kind of definition of *nationa* along ancient lines of race, ethnicity, or even common cultural heritage quite meaningless.

This does not imply homogenization. Indeed, the levels of economic, political, and cultural diversification are astounding compared to your times. But this occurs within an intricate system of mutual self-support and non-exclusionary practices across the whole *nationa*.

Some rough sort of answer has therefore been devised to the dilemma of constructing on the one hand a well-ordered system for the purposes of guaranteeing adequate life-chances for all, while, on the other hand, allowing for that kind of chaotic disorder that provides the seedbed for creative interactions and personal self-realization.

What is interesting about the *hearths* and *neighborhoods*, for example, is the degree to which they achieve a high degree of self-sufficiency and are themselves centers for the creation of cultural and lifestyle diversity. They are centers of intense sociality and cultural experimentation, places where the art of after-dinner conversation, of musical performances and poetry readings, of 'spirit talk' and story telling, are so tenderly cultivated as to make them sites of continuous social engagement (not all of it harmonious by the way!).

They are places where people who want to be different can express that want with the greatest freedom.

Their general manner of organization is symbolized by the physical outlay. The nuclear family home that dominated urban form in your countries and times had to be displaced by some more collective arrangement (the prototype turned out to be some of the structures set up by the polygamous Mormons in the American West in the late twentieth century). Whole city blocks have been converted. Entrances have been punched through dividing walls among the row houses or, in the case of detached housing and sprawling suburbs, walkways and infill rooms have been constructed between the already existing structures to link them into a continuous unit of high-density habitation (liberating some formerly sprawling suburban tracts for intensive cultivation). Larger spaces have been carved out as the common kitchens and eating halls but everyone has their own private room equipped with some basic equipment and within which they are free to use their space as they wish.

Several *hearths* are linked together around a *neighborhood* center that houses both general educational and health care functions. In some instances we have adapted older town and even city centers to these tasks (though our virulent opposition to any structures higher than four or at most seven storeys meant major transformations in urban design in what used to be called the West). Elsewhere, particular traditional forms and ways of life have been adapted as need be to the new circumstances. *Neighborhoods* are also points of intense social interaction and entertainment (the *edilia*'s centralized store of videos and recordings can here be tapped into at will).

The sheltered areas within the reorganized city blocks are mostly converted into walled gardens – with a few appropriate play-spaces for children and restful arbors for adults – in which all manner of intensive cultivation occurs (producing fruits and vegetables galore) supplemented by glass houses and hydroponic cultivation systems that guarantee a year-round supply of everything from salad vegetables to excellent high quality marijuana (the main recreational drug of choice).

Urban agriculture and gardening is a prominent feature (some derelict land in New York City was used this way in your day). This has both a social as well as an economic significance since many people evidently take pleasure in such activities. On pleasant days the gardens become a venue for a lot of socializing and 'spirit talk.'

Composting of organic wastes is combined with an adaptation of an ancient Chinese system of night-soil circulation (a triumph of bio-chemical engineering) so that nutrients are re-cycled on both a *neighborhood* and *edilia* basis. You doubtless remember Victor Hugo's comment that 'the history of civilization lies in its sewers' – well, we present our system as evidence of a society well on the way to forming a radically different kind of civilization!

The roofs of the habitations are adorned with solar panels and small wind sails (the effect is somewhat Heath-Robinsonish and would probably not be aesthetically pleasing to you). Powerful batteries store energy in basement areas supplemented by an elaborate system of fuel cells (an innovation perfected by the military). A variety of other local sources of energy are likewise mobilized.

Energy reliance on the outside is generally regarded as debilitating and degrading. Each *neighborhood* has its buried reserve of oil for emergencies. But it is a matter of pride not to use it (replenishment is costly). People often prefer to bundle up (or cuddle up!) in cold weather rather than use up that oil.

Everyone partakes of the tasks within the *hearth*. Cooking, cleaning, and all the other chores are divided up on a rota. Most *neighborhoods* have individuals trained to undertake the maintenance necessary for the physical structure and its electronic systems.

In addition, each *hearth* specializes in some particular kind of production (baking, brewing, sewing, dress and shirt-making, pasta-making, sauces, conserves, etc.) which it exchanges with other *hearths*. (Your expertise in baking and conserving would be put to good use!) The communications system is here crucial. Orders can be placed electronically and surpluses advertised on bulletin boards. Work and exchanges can then be organized on an efficient and waste-minimizing basis.

Particular *hearths* often use their labor credits (these will shortly be explained) to build a specialized stock of raw materials (dried beans, rice, flour, sugar, coffee, cloth, and thread, or whatever) which can be traded to other *hearths* for other produce. A collective store of surplus produce is dispersed among the *hearths* so that the *edilia* is well stocked with several months of basic non-perishable provisions.

Labor is exchanged among the *hearths*. For example, major construction or renovation projects within the *neighborhood* are undertaken by workers of different skills from various *hearths*, though in some instances labor from other *neighborhoods* is 'imported' in exchange for reciprocal engagements at some later date. Computerized balances of labor credits provide a rough accounting measure to ensure fair exchange.

Hearths also acquire fame and reputation according to the generosity and sophistication of the hospitality they offer. The reciprocal exchanges that take place between them in this way have become a vital aspect of social and political life. Competition with respect to generosity towards others is an important value.

The *hearths* might seem to you oppressive to the individual. They are not, of course, conducive to the more blatant forms of individualism to which you are accustomed. But individuals can change *hearths* if they wish. Since the *hearths* are so different in character (they vary according to food preparation styles, musical tastes, cultural expression, traditions, gender mixes) the individual has even more choice of *hearth* type than New Yorkers of your generation had with respect to ethnic restaurants.

The only restraint is that changes should occur on an annual basis and at a given time (the first week of November is the general date of removal). Notice of one month of intention to move is required so that vacancies can be advertised on the computer bulletin board and hopefully filled. Most people choose not to move, but individuals who feel oppressed have the opportunity to do so and some restless individuals (you would probably be among them) never settle for long in any one *hearth*.

This arrangement is supplemented by another important innovation – the institution of 'sabbaticums'. Every seventh year (after the age of seventeen) each and every individual is entitled to spend a year elsewhere (almost anywhere in the world).

Such individuals must commit themselves as full participants of whatever *hearth* they become attached to. But sabbaticums provide a remarkable opportunity to learn and to explore the world in a different way. Many choose not to avail themselves of the opportunity (more than, say, once in a lifetime), but an almost equal number regularly take their sabbaticums with profit and pleasure, sometimes seizing the opportunity to retrain themselves in quite different skills.

Individuals also have their own private resource budget that gives them limited rights to exchange goods and services on their own account outside of the framework of the *hearth* economy. This is a residual from the many local economic trading systems (LETS) that sprang up at the time of the collapse (you even had a few of them in your day).

It amounts to an extensive computerized bartering network. Individuals have to earn their points by providing particular goods and services to others and can then use those points to acquire things or services they need for themselves. Individuals enter into this system at age seventeen when each person receives an initial social endowment, set by the wealth of the *edilia*, which gives them an initial number of points to begin to trade. Inheritance is thereby made a collective rather than a personal and individual good.

Since much of the basic organization of sustenance had been taken over by the collective activities of the *hearth*, individual exchanges have declined somewhat in significance. But there are all sorts of items, such as collectibles and 'vanity' objects, that trade in this way. The appetite for higgling and haggling, for truck and barter, is thereby satisfied. The innumerable 'flea markets' and other informal markets scattered throughout every *edilia* testify to the importance of this activity. It must be said, however, that 'going to market' is now regarded more as an occasion for sociality than serious trade. Objects are often traded more for the opportunity of conversation and social contact than for any thought of economic advantage.

The most prevalent form of this activity (and probably most shocking to your prurient eyes) is the exchange of sexual favors, a practice that deserves a somewhat fuller and franker discussion. In effect, all those 'personal' ads and 'dating networks' that littered the media in your times have been converted into a computerized system for organizing sexual exchanges (were they really ever anything more?).

We fully recognize what should long ago have been obvious: that the relationship between sexual activity on the one hand and the organization of parenting on the other is entirely accidental.

The last-ditch efforts by theocratic powers to keep the family alive as the foundation of the social order were increasingly seen as an exercise in social control through the tight control over the desires, dispositions, and even functions of individual bodies. The demise of theocracy coincided with the collapse of these controls.

There then followed an intensive debate on how sexual life and parenting activities might be so organized as to displace the dysfunctional and ancient structures of the family (an institution which occasionally worked well in your days but was for the most part the center of violence, abuse, alienation, and, worst of all, neglect of the real needs of children to be raised in secure, loving, and supportive environments).

The solution we have arrived at looks something like this. What are called *pradashas* comprise a number of individuals who bond together in an irrevocable contract for the purpose of raising children. The unit can be of any size but the minimum is six adults and the average is between eight or nine. Instances can be found where the *pradasha* is all male or all female but usually there is some mix of sexes. 'Adjuncts' can also be participants in parenting though they are not bound to the unit in any irrevocable way. They act as grandparents and aunts and uncles once did and can be called upon for help when necessary. The whole is constructed rather like the large extended family of yore, though it is now based on voluntary bonding rather than blood relations.

The formation of a *pradasha* takes a good deal of preparatory negotiation. It brings together people who take great pleasure in the having and raising of children and creates a supportive environment for that activity. The sole focus on child rearing will, we feel sure, shortly pay off, raising children whose psychic make-up and attitudes are totally different from those you are familiar with.

Relations based on affection, love, and respect are fundamental to the well-being of the *pradasha* and this includes sexual relations. But individuals are entirely free to enjoy sexual relations with others (of any sort). To this end an extensive computerized market for the exchange of sexual services exists in which individuals earn points by providing sexual favors for others and spend points on procuring sexual favors from others.

This freedom is accompanied by certain measures that to you would seem draconian. By 2005 a contraceptive system for men had finally been designed and it became possible to insert a small capsule within the blood stream to last for one year.

All men who wish to be placed on the dating network have to submit to this medical procedure from pubescence onwards. Furthermore, careful monitoring of sexual health (sexually transmitted diseases in particular) is now mandatory. The sexual health of any prospective partner is coded into the dating network and available for inspection.

This would, of course, be anathema to your mode of thinking but if one considers the incredible freedoms it confers in other ways then the thought should be more palatable. Indeed, it seems to us that your intense preoccupation with privacy in such matters had more to do with controlling sexual behaviors through fear than with the true protection of individual rights.

Sexual exchanges of this sort have some unintended benefits as well as problems. On the positive side the points system leads to a much greater equality between the sexes with respect to sexual practices. Men, for example, in order to gain points, must build a reputation with respect to the satisfaction of women and the sexuality of the latter has become much more dominant as a result.

The women's movement – many of whose members were forced to participate in the sex trade in order to make ends meet in your times – actually took the lead in setting up this more egalitarian system during its movement for disarmament. They also took the lead, largely out of necessity, in the design of collective systems of child care that later evolved into the *pradashas*.

The exploration of all manner of sexual relations also means the disappearance of categories like 'heterosexual,' 'gay,' and 'queer' since no one particularly holds to such identities anymore but freely roams among sexual practices as they see fit.

The biggest danger is that of sexual obsession – the inability to treat sexuality as a source of pleasure and the desire to use it for possession of the other or for the total merging of personalities. This has proven one of the deepest problems and it has taken a lot of hard work and careful counseling to limit the potential damage that comes from such bad habits.

The balance we now strike between sexual exchanges and parenting works very well. Children are nurtured in the right kind of way, full of love and attention, and the joy of so doing is as widespread as anyone could want. They are no longer viewed as property, as possessions, and their development occurs outside of those awful blandishments of the commodity economy and the struggle for personal advantage that so darkened their lives in your day. The world of parenting is protected from the potentially disruptive and destructive effects of the pursuit and liberation of sexual desires (making all the divorces, extra-marital activities, and prurient public morality of your times a thing of the past).

Parenting is also immune to the damage wrought in your times by crushing demands of a labor market oriented solely to the production of profit rather than to public wants and needs.

This illustrates, however, a more general principle. There can never be a condition of total freedom in any society. Some sort of balance has always to be struck between individual rights and the personal pursuit of desires on the one hand and collective rights, rules, and obligations on the other. The revolution of 2020 simply changed the balance around relative to that which you now know. It liberates individualism and the pursuit of self-realization in certain directions because it does not fear to constrain them in others.

This shift is nowhere more apparent than in the conversion of the system of personal identifications set up initially by the credit agencies and pressed home by the military. Libertarians wanted to abolish it entirely.

But the women argued it was not a bad thing to be able to know instantly by scanning who one is dealing with. Someone remembered a proposal that democracy might best be served by opening up the data banks to everyone rather than trying to do without them or, even worse, pretending they did not exist. Everyone now knows about everyone else what all credit and government agencies knew in your times (and a good deal more besides). That information is instantaneously scanned between one person and another, so nobody can use it for privileged or authoritarian purposes.

The loss of a presumed (but in practice mainly fictitious) privacy has to be accepted but the gain in individual security is immense. The identity of any perpetrator of some violent act, for example, is immediately known. *Hearths* and *neighborhoods* can become open to all, since all outsiders are easily identified. All social spaces are opened up in hitherto inconceivable ways – security no longer depends upon walls, doors, locks, fences, electronic barriers. It rests on the simple knowledge, available to all, as to who is present where.

Take another example. Free transportation is available to everyone. The efficient and ecologically sound transportation systems devised by the military were put to excellent use to facilitate exchange and movement around the world.

Locally, small electric cars (designed not to travel more than twenty miles and to move no faster than twenty miles an hour) and bicycles are located at the edge of every *neighborhood*. The insertion of a key-card allows any adult to command any vehicle and to plug it back in on arrival at one's destination (the vital key-card is released only on satisfactory reconnection and abuse of any equipment can quickly be tracked down).

Locomotion may be slow and restricted but it is, like sex, free and safe.

This also illustrates another important feature of our society. It has generally slowed down rather than speeded up. It has also become notably much quieter: the violent levels of noise pollution that were such a bane in your time have largely disappeared. Of course, in this as in many other things, considerable variation exists between *neighborhoods* and *edilias* – some of the latter, much beloved by some of the younger generation, are marked by a frenzied style of life and loud and boisterous festivals. Some young people also exercise their love of movement by roller-blading at fantastic (and occasionally, we regret to report, fatal) speeds.

Technological innovation in your times had never really been about lightening the load of labor or making life easier for anyone: it was about gaining profits and pressuring whole populations into behaviors that made them like cyborgs (human appendages of machines both in the workplace and, even more insidiously, in the home). It forced levels of stress to supremely high levels through its relentless pursuit of speed-up and intensification. Emancipation from want, pressures, and needs was not its aim. The prospects for humanizing technical relations were foreclosed. That is how we see it now.

We hold that technology must make life simpler rather than more complicated. Technological change has not stopped. Indeed, there is great emphasis upon it (particularly in areas such as electronics, genetic medicine, which we see as the ultimate cure for many diseases, and materials science) but the rules that govern its application are quite stringent.

Committees evaluate new technologies in the following terms. They must:
1. lighten the load of labor;
2. be environmentally friendly if not beneficial (generating wastes that are non-toxic and easily reused, for example);
3. be less rather than more complicated and therefore more easily used and maintained by all; and
4. be consistent with the ideal that labor is as much a social as a technical activity (technologies that isolate people are less favored than those that bring them together).

Finally, technology should work to the benefit of all, with special emphasis upon the least privileged.

These requirements are not always compatible (as you will doubtless quickly object). So the committees are often hard put to reach a decision. We cannot claim we always judge right. But uncertainty of this sort never hurt anyone and we actually like the challenge.

Another trenchant symbol of change is calendar reform. This had been tried several times before (most notably during the French Revolution) but it never succeeded until now.

The abolition of weekends will doubtless shock you. This happened because of the religious significance of Saturdays and Sundays and the abusive ways theocratic power manipulated whole populations through calls to submission and worship at these times.

We now have a five-day week with six weeks to each month which leaves five days over (six in a leap year) which we insert between June and July as 'Festival Days' – this is an occasion of wild celebration (for which *edilias* and *neighborhoods* prepare assiduously throughout the year). It makes events like your Mardi Gras and the Rio Carnival pale in comparison.

But back to more serious matters! Each adult person is expected to 'work' three of the five days in a week with five hours 'work' a day. Flexibility is built in so individuals can work two days of seven and a half hours, or whatever. This means a workload of around ninety hours per month and, with one month off a year plus a sabbaticum every seventh year, the formal working time of each person over a lifetime is much reduced.

Hearths can also designate ten other days as 'days of ritual.' Most *hearths* use at least four of their days to participate in *neighborhood* and *edilia* celebrations (rather like your street festivals), but the rest remain particular to each *hearth*.

All sorts of occasions are chosen (though it is notable that older religious and mythological rituals – like the day of the dead – are often preserved). The manner of celebration varies from quiet inner contemplation to boisterous 'guest days' where outsiders are bidden to visit, eat, drink, perform, or whatever.

There is another interesting adjustment to the work system. Half the collective labor of a *hearth* is normally allocated to activities organized by the *edilia* or *regiona* in return for needed materials, rights, and services (such as free transportation, machinery and equipment, building materials, etc.). The other half is given over to activities within the *hearth* destined for exchange with other *hearths* (the baking, brewing, sewing, etc. already mentioned).

One of the most important discussions within the *hearth* (and sometimes a cause of bitter dissension) is how to vary these proportions. The supply of labor hours to the *edilia* or *regiona* can be cut down if the *hearth* decides upon lifestyle changes or internal production mechanisms to make it less dependent upon external exchanges.

If the *hearth* wants materials from the *regiona*'s factories or increased rights to transportation, it must send so many more units of labor to the *regiona* in order to get them. But the number of labor credits needed to procure products or services from the *edilia* or *regiona* varies. When labor is lacking at the *edilia* or *regiona* level, it takes a lot more labor credits to procure materials and services from them.

A curious kind of labor market arises. Exchange ratios (measured in labor credits) between *edilias*, *regionas*, and *hearths* fluctuate on a monthly basis.

This system results in occasional instabilities but it generally runs smoothly enough. Most *hearths* provide a steady supply of labor to both *regionas* and *edilias* in return for a steady supply of credits to be used to procure whatever they need.

The occasional shortages that result are an occasion for loud complaints. But they also have the positive effect of encouraging a lot of recycling (everything from nails to plastics and paper) and reminding everyone that the bad old days (typical of your times) of the throw-away society and instantaneous obsolescence must not return. In any case, such scarcity as exists has the virtue of being equally shared instead of being foisted upon the least fortunate (as happened in your day).

Perhaps the most beautiful aspect of this system, however, is the gradual dissolution of the boundary between work and play. While it is commonly recognized that some formal accounting system is needed for society to function at all, the fact is that active people are active because they enjoy being active and much of that activity is now channeled into pleasurable but productive work.

What once passed as hobbies have became part of production and much of production is organized as if it were a hobby. Gardening and orchard management, education of young children, care of the environment, doing carpentry and small odd jobs of improvement, even major projects to extend or refurbish buildings, as well as cooking and experimentation with cultural forms (painting, mathematics, music, poetry, etc.) are all organized in a way that has nothing to do with 'formal' work requirements and everything to do with the pursuit of a satisfying social life. The *hearths* are forever organizing projects for their own amusement and self-betterment.

In this way the number of formal hours given over to children's education has been much reduced. Children daily accompany adults (most, though not all, of whom enjoy the experience) into the gardens and orchards, into the hot-houses or down to the fishponds, into the workshops, or wherever. They do practical work while learning about botany, biology, principles of agronomy, mechanical arts, and the like.

There are no formal places of worship anymore. The churches are converted to other uses – the smaller ones to communal living spaces, others into *neighborhood* centers and still others into large recreation halls (gymnastics is a much favored activity), or places where concerts, theatrical events, poetry readings, music competitions, and the like can take place. The beauty of such spaces is thereby preserved (you who often listen to musical events in some ancient cathedral will appreciate what we mean).

Music is much loved but it assumes an incredible variety of forms. Remarkably, mathematics and poetry have become connected to it and each so enhances the other as to create a general conception of the poetry of the universe to which everyone in one way or another subscribes. These three have become the focus of the greatest public celebrations.

The collapse of formal religions does not mean any loss of spirituality. Indeed, people appreciate and admire all forms of 'spirit talk.' And they still read and venerate the religious texts as beautiful stories and as morality tales that contain intense spiritual insights as well as practical lessons.

What we call 'spirit talk' is not confined to preachers or learned individuals. It is open to all. When the feeling strikes, individuals communicate their ideas in the home, in the workplace, in the streets, or in any number of public places.

You would doubtless look upon this with horror, imagining that the whole world has been taken over by those street corner ravers you often encounter in New York City. But now this practice has become integral to a way of life that centers upon the convergence of spiritual powers and rational orderings. Individuals can explore with childlike curiosity the realms of thoughts, feelings, and dreams. And they can do so in an atmosphere of the greatest spontaneity.

It may sound as if we lack contentiousness. But that is simply not true. Conflicts and disputes (not only between the spirit talkers who often argue vehemently) are viewed positively rather than suppressed. The dialectics of contention is widely viewed as fundamental to self-realization and social change. There are, however, some remarkable differences in how disputes are formulated, confronted, and resolved.

To begin with, the profession of 'lawyer' has entirely disappeared (a historical event that most of your generation would have devoutly wished even as they were drowning in their own litigiousness). In retrospect we see the legal community as a prime culprit in hastening the earlier descent into barbarism.

But the traditions of law (like those of religion) have been preserved because they are widely recognized as a crucial preparation for a civilized social life. In the past, however, they always remained just that: a preparation for something that never arrived. So we preserve the traditions of law but dispense with the professions of lawyer and judge.

This attitude carries over into other areas. The universities, for example, have been disbanded. Taken over and administered entirely by large corporate powers by the early years of the twentieth century, they had by 2010 become centers for corporate/military research or for the privileged training of a self-replicating corporate/state elite. The only kind of traditional scholarship tolerated was an obfuscating academicism specifically designed (or so it seemed) to take all pleasure out of learning and to prevent the formation or communication of significant ideas.

But love of learning did not disappear. Now freed of its professiona-lization it has undergone a remarkable resurgence. Individuals pursue their love of literature, poetry, mathematics, history, geography, science, and the arts (mechanical and technical as well as traditional) in all sorts of ways. They do so in an atmosphere of intense pleasure, enjoyment, and disputation, though always in a manner compatible with their other duties (as carpenters, graphic designers, cooks, or whatever). The supreme art of translation is highly venerated and valued.

Many choose to take a sabbaticum (or even to re-locate full-time) in a particular *neighborhood* or *edilia* where certain groups of people have come together to share a common love of some sort of learning. Others use it to learn the humble craft of translation through total immersion in the ways of living of others.

Young people, who in their seventeenth year are required to spend at least a year away from their *pradasha*, often go to a place where some noted savants have gathered to study, for example, sciences, mathematics, law, religion, totemic systems, or the great literatures of India, China, and Europe. While most information can now be stored electonically, in some *edilias* many old books and manuscripts can be found, along with local techniques of book production that sustain the ancient pleasure that comes with curling up in a corner with a good book.

The abolition of most forms of private property and the transition to property held in common eliminates much of the legal contention that dominated your world, but disputes are frequent and occasionally severe. Some rough rules and customs have arisen for conflict resolution in dif-ferent arenas.

Disputes that escalate into anger and strife within *pradashas*, for example, are regarded very seriously. The wisdom of experienced indivi-duals within the *neighborhood* or *edilia* is quickly mobilized to dampen such strife.

The whole approach to retribution, justice, and punishment has chan-ged. While transgressions against others (violence in particular) have greatly diminished because the causative seedbed has largely been ploughed under, cases still occur. These are intially viewed as disharmo-nies within the person committing the act. The response (as used to be the case with the Navaho) is to find means to unravel the upset and restore the harmony.

Persistent offences can lead to sanctions, such as the withdrawal of trading privileges. In extreme cases of persistent offences, banishment to the 'dangerous lands' is possible (a decision made by the *edilia* council).

This means assignment to areas contaminated during the revolutionary wars where life is still perilous. Such areas cannot be abandoned because the toxicities and diseases fomented there pose serious threats. Commissions and workparties drawn from the *nationas* here collaborate, and the need for a non-volunteer workforce is partly met by individuals banished from their *edilia* in order to work off some punishment for persistent crime and violence.

Disputes between *edilias* or *regionas* are worked out by negotiating committees. At the *regiona* level, most disputes are over trade relations which, by universal agreement, have become based on equality, non-coercion and reciprocity (we have made Adam Smith's theory of perfected markets practical and real, though largely through bilateral agreements). Disagreements on trade matters are routinely disposed of without much fuss.

Global Councils with well-publicized advisory functions exist, however, to consider some rather more difficult issues such as the paths of technological change, production formats, environmental problems, management of those resources (such as biodiversity and the oceans) considered as part of the global commons, and some matters pertaining to disaster relief, population relocations, and the like.

These councils are supposed to alert everyone at all levels, from the *nationas* to the *hearths*, to the existence of common global problems that perhaps require local solutions. Established first in the brief period of World Governance, these councils are now largely advisory and consultative. But they continue to play a highly influential role in formulating agreements between *nationas*.

The system of political representation is extraordinarily simple. We will be brief so as not to bore you too much.

Each *hearth* elects a representative to work on a *neighborhood* council for a three-year non-renewable term. Each *neighborhood* elects a person to serve on an *edilia* council for a three-year non-renewable term, and so on up the scale to the *nationa* which appoints representatives to the Global Councils.

Each level of government above the *neighborhood* can second personnel (no more than 5 per cent of the population) for no more than ten years to work in a technical/administrative secretariat. In such a capacity, individuals can work on scientific/technological commissions or upon research and development as well as upon the improvement of systems of allocation/distribution via the computerized bidding systems and bulletin boards.

But it is a strict rule that all such personnel must attach themselves to a particular *hearth* where they participate in activities in a normal manner (the labor credits they receive for their activities are much appreciated in the *hearth*). The dispersal of such personnel through many *hearths* (and the periodic geographical shift of commissions and secretariats from one *edilia* to another) ensures strong contact between persons in the *hearths* and those operating at higher levels of governance. We thus avoid that practice, so damaging in your time, of allowing a privileged bureaucratic elite to ghettoize itself and get out of touch with the daily lives, wants, needs, and desires of the people.

Some production activities are orchestrated via the *nationa*. We speak here of things like electronics, silicon chips, metal working, engineering, transportation, communications systems, and textile fiber production. Such sectors are highly automated and require little labor. They are usually organized to combine economies of scale with economies of scope and are able to switch gears quickly from one category of product to another (e.g. silicon chips for different purposes or electronic equipment of different sorts).

There is, therefore, a strong element of what you once condemned as 'undemocratic centralized planning' (dismissed as 'socialistic' or 'communistic'). Much of this planning is to be found at the *regiona* level and it plays a key role in mixing the need for order in production with the desire for localized disorder as a seedbed for cultural renewal.

Agriculture is likewise divided between large-scale low-labor input systems of production for grains, raw materials, beans, and legumes, and the labor-intensive activities connected with the fishponds, the gardens, and the hydroponic cultivation systems. Management Committees at the *regiona* level struggle to get the balance right between these two sorts of agriculture in an attempt to construct long-term sustainability, self-sufficiency, and sociability.

One excellent side effect has been the shift in preferences towards a much healthier diet based in grains, beans, legumes, vegetables, nuts, and fruits. The meat content (always a preserve of the affluent and a terribly inefficient means to feed people) has been much reduced. This has permitted the abolition of obnoxious and degrading practices developed in your times for the production of beef, veal, and chicken.

Some of the personal habits and values established after the revolution might seem strange or even objectionable to you. Many people became used to somewhat Spartan lifestyles consistent with local self-sufficiency. They also acquired a rather hardened mental attitude to hurt and injury because situations demanded it, even if every effort was made to attend to the physical comfort of someone in pain.

This general attitude has helped avert any return to the self-indulgent hypochondria (bordering on mass hysteria) that characterized the pill-popping elite classes in your times (aren't you on Prozac too?). It also prevents the medical establishment from pandering (as it so patently used to do) to imagined as opposed to real ills. And it has forced psychoanalysis to return to its roots and accept that the treatment of mental ills lies in the art of intimate, excellent, and probing conversations. A professionalized business has become a general art form.

We consider our medical care highly sophisticated. Every *hearth* has at least two people knowledgable about minor problems and from *neighborhood* to *edilia* to *regiona* some sort of hierarchy of information flow and of facilities exists. Many (though not all) of the large-scale hospitals have been disbanded in favor of *neighborhood* caring units.

Health care is oriented more to prevention than to cure. The demand for palliative drugs is much diminished (your own drug industry typically opposed prevention and even cures in order to perpetuate its profits through palliative drug dependency). The other interesting feature is that medical care, understood as an art as well as a science of healing, frequently diverges in qualities and style from one place to another.

While it is generally accepted, for example, that dying with dignity is a right, the manner of so doing varies greatly from the relatively private and quiet to the very social and even boisterous. Death is not feared but understood as integral to life and it is simply not the style to go to extraordinary lengths to prevent it at any cost (as was the case for the privileged elite of your times).

Death is seen as a moment of intense sadness and celebration, the moment of eternal return of the human spirit to its origins and the moment of transmission of all that has been accomplished through a life to another generation. It is the moment for everyone to reflect and take stock, to come to terms with their own life and death, and to recommit themselves to activities and relationships that will be worthy of transmission to future generations.

It is perhaps for this reason that many of us now believe that the spirits of the dead continue to circulate among us, always.

But we must have done with this tale! Out of the many other things that can be said we address only those of the greatest import.

Perhaps the hardest of all things to convey (particularly to a whole generation of sceptics and cynics of your sort) is the spirit that pervades this new social order. It is not as if the will to power, the excitement of performance, the pursuit of passions, or the adventurous curiosity of individuals and groups have been put under wraps. Quite the contrary. All of these elements flourish even as they are channeled down different paths.

It has been, above all, the revolution in values that has made the difference. And in this regard the most important point has been left to last. The absence of money from our world is startling. On this point we have nothing original to say, for it was all so well stated by Sir Thomas

More (in 1516, no less!). You perhaps remember it. Existing society, he argued, is nothing but:

> [A] conspiracy of the rich to advance their own interests under the pretext of organizing society. They think up all sorts of tricks and dodges, first for keeping their ill-gotten gains, and then for exploiting the poor by buying their labour as cheaply as possible. Once the rich have decided that these tricks and dodges shall be officially recognised by society – which includes the poor as well as the rich – they acquire the force of law. Thus an unscrupulous minority is led by its insatiable greed to monopolize what would have been enough to supply the needs of the whole population.

But in Utopia:

> [W]ith the simultaneous abolition of money and the passion for money, how many other social problems have been solved, how many crimes eradicated! For obviously the end of money means the end of all those types of criminal behavior ... And the moment money goes, you can say good-bye to fear, tension, anxiety, overwork and sleepless nights. Why, even poverty itself, the one problem that has always seemed to need money for its solution, would promptly disappear if money ceased to exist.

Computerized exchange transactions of the sort pioneered in electronic banking in your times permitted the abolition of the very money exchanges that the system was designed to facilitate. It is now possible to engage in the multiple bartering of services of all sorts, from sexual favors to pots and pans, without using money at all. The whole social world is now turned upside down in such a way that the exchange of meaningful uses as opposed to the senseless pursuit of money power has become the dominant motif of the social order.

The grand debate that we now have is over what is 'meaningful' about a particular use. And here, the huge unanswered question mark that still animates innumerable passions is simply this: 'what can the true nature of human nature truly become ...'

. . .

I awoke in a cold sweat. Had I had a dream or a nightmare? I prized my eyes open and peered out of the window. I was still in the Baltimore of 1998. But I was unsure whether to be reassured or distressed by the fact.

The dream stayed with me for much of the day. The general picture I was left with was down-to-earth, commonsensical and in some ways very attractive. But there were many elements that left me anxious and nervous the more I thought about them.

Imagine a world with no banks or insurance companies to run our lives, no multinational companies, no lawyers, stockbrokers, no vast bureaucracies, no professors of this or that, no military apparatus, no elaborate forms of law enforcement.

Imagine all those workers freed from their subservience to the pathetic and parasitic activities to which they are now bound. Imagine them cut loose to work on productive tasks in a world where an ecologically friendly technology requires no more than a few hours work a day to take care of basic needs.

Imagine the loss of the frenetic pace of contemporary life and the transformation of those moments of pure enjoyment we now perforce snatch in between pressing obligations into hours of plenitude.

Above all, imagine a world of respectful equality, not only of talents or achievements, but of conditions of life and of life chances – a world, in short, where the ugly habit of shifting the burden of one's support onto the shoulders of others has disappeared.

Imagine a world in which the pursuit of pecuniary advantage no longer matters and the glitter of all that is gold has lost its allure.

The vision was in one sense exhilarating. But the loss of all those usual props to daily life was also terrifying.

I relaxed with my cappuccino. The stock market, after a summer of ups and downs, was at a new high. And I took comfort in the fact that everything I had dreamed was obviously so outrageously and outlandishly foreign to our contemporary ways as to be outside the realm of any possibility. Any discussion on this set of possibilities was bound to get a bad press, I said to myself, 'and rightly so.' It was, after all, a very undialectical tale powered by exactly that kind of implausible apocalyptic scenario I so disliked.

I walk the streets of Baltimore.

Massive monuments to the rich tower oppressively around me. An elaborate state-subsidized welfare system supports hotels, corporations, up-scale condominiums, football and baseball stadiums, convention centers, elite medical institutions, and the like. The affluent build a system of private schools, universities, and medical establishments that are the best in the nation while the mass of the excluded population drown in the miasma of a public sector that is so busy subsidizing the rich that it cannot achieve even elementary standards of performance for the mass of the population.

The suburbs boom with unecological sprawl while forty thousand vacant houses in the city disintegrate and decay. A filthy ozone cloud hovers over the city on warm summer days. Forty thousand IV drug users roam the streets; the soup kitchens are pressed to full capacity (as are the jails); the food banks for the poor have run out; and the Malthusian specters of death, famine, disease, and the 'the war of all against all' hang like a pall over the streets of the city.

Where is that order of unity, friendliness, and justice which Howard invoked? If my dream had had some aspects of a nightmare then wasn't this reality every bit as nightmarish?

I had always thought that the purpose of More's *Utopia* was not to provide a blueprint for some future but to hold up for inspection the ridiculous waste and foolishness of his times, to insist that things could and must be better.

And I recalled how Bellamy's hero returns to the Boston of 1888 to find it even more appalling than he had ever thought, at the same time as he is mocked and ostracized for talking of alternatives. His contemporary reality turned out to be his nightmare. Is it not just as surely ours?

If, as most of us believe, we have the power to shape the world according to our visions and desires then how come we have collectively made such a mess of it? Our social and physical world can and must be made, re-made, and, if that goes awry, re-made yet again. Where to begin and what is to be done are the key questions.

Looking Backward, Bellamy noted, was written, like now, at a moment 'portentous of great changes.' It was also written

> in the belief that the Golden Age lies before us and not behind us, and is not far away. Our children will surely see it, and we, too, who are already men and women, if we deserve it by our faith and by our works.

And when that Golden Age arrives we may finally hope 'to say good-bye to fear, tension, anxiety, overwork, and sleepless nights.'

Bibliography

Abers, R. 1998, 'Learning democratic practice: Distributing government resources through popular participation in Porto Alegre, Brazil,' in Douglass, M., and Friedmann, J. (eds), *Cities for Citizens*, New York, 39–66.

Alston, P. (ed.), 1992, *The United Nations and Human Rights: A Critical Appraisal*, Oxford.

American Anthropological Association, 1947, 'Statement on human rights,' *American Anthropologist*, 49, 539–43.

Amin, S. 1974, *Accumulation on a World Scale*, New York.

Arrighi, G. 1994, *The Long Twentieth Century*, London.

Avineri, S. 1972, *Hegel's Theory of the Modern State*, London.

Bacon, F. 1901 edition, 'New Atlantis,' in *Ideal Commonwealths*, London, 103–7.

Bakhtin, M. 1981, *The Dialogic Imagination*, Austin, Texas.

Beck, U. 1992, *Risk Society: Towards a New Modernity*, London.

Bellamy, E. 1888, *Looking Backward*, New York.

Benjamin, M. 1998, 'Time for a living wage, around the world,' *Global Exchanges*, 36, 1–5.

Benjamin, W. 1969, *Illuminations*, New York.

Benton, T. 1993, *Natural Relations: Ecology, Animal Rights and Social Justice*, London.

Birch, C., and Cobb, J. 1981, *The Liberation of Life: From the Cell to the Community*, Cambridge.

Blakely, E. 1997, *Fortress America: Gated Communities in the United States*, Cambridge, Mass.

Blaut, J. 1977, 'Where was capitalism born?' in Peet, R. (ed.), *Radical Geography*, Chicago, 95–111.

Blaut, J. 1993, *The Colonizer's Model of the World*, New York.

Bloch, E. 1986, *The Principle of Hope* (4 vols), Oxford.

Bloch, E. 1988, *The Utopian Function of Art and Literature*, Cambridge, Mass.

Bohm, D. 1983, *Wholeness and the Implicate Order*, London.

Borja, J., and Castells, M. 1997, *Local and Global: Management of Cities in the Information Age*, London.

Bourdieu, P. 1984, *Distinction: A Social Critique of the Judgement of Taste*, London.

Brooks, R. 1992, 'Maggie's man: We were wrong,' *The Observer*, Sunday June 21st, 1992, 21.

Butler, J. 1993, *Bodies That Matter: On the Discursive Limits of 'Sex'*, New York.

Butler, J. 1998, 'Merely cultural,' *New Left Review*, 227, 1998, 33–44.

Campanella, T. 1901 edition, 'City of the Sun,' in *Ideal Commonwealths*, London.

Capra, F. 1996, *The Web of Life*, New York.

Cardoso, F., and Faletto, E. 1979, *Dependency and Development in Latin America*, Berkeley.

Castells, M. 1996, *The Rise of the Network Society*, Oxford.

Castells, M. 1997, *The Power of Identity*, Oxford.

Castells, M. 1998, *End of Millenium*, Oxford.

Chevrier, J.-F. 1997, *The Year 1967: From Art Objects to Public Things*, Barcelona.

Clark, L. 1991, *Blake, Kierkegaard, and the Spectre of the Dialectic*, Cambridge.

Deleuze, G., and Guattari, F. 1984, *Anti-Oedipus: Capitalism and Schizophrenia*, London.

Derrida, J. 1994, *Given Time: 1. Counterfeit Money*, Chicago.

Derrida, J. 1994, *Specters of Marx*, London.

Dickens, C. 1961 edition, *Hard Times*, New York.

Douglass, M., and Friedmann, J. (eds), 1998, *Cities for Citizens*, New York.

Duany, A. 1997, 'Urban or suburban?' *Harvard Design Magazine*, Winter/Spring 1997, 47–63.

Eagleton, T. 1997, 'Spaced out,' *London Review of Books*, April 24th, 1997, 22–3.

Edsall, T. 1984, *The New Politics of Inequality*, New York.

Elias, N. 1978, *The Civilising Process: The History of Manners*, Oxford.

Emmanuel, A. 1972, *Unequal Exchange: A Study of the Imperialism of Trade*, New York.

Etzioni, A. 1997, 'Community watch,' *The Guardian*, June 28th, 1997, 9.

Fernandez-Kelly, P. 1994, 'Towanda's triumph: Social and cultural capital in the transition to adulthood in the urban ghetto,' *International Journal of Urban and Regional Research*, 18, 88–111.

Fishman, R. 1982, *Urban Utopias in the Twentieth Century*, Cambridge, Mass.

Fishman, R. 1989, *Bourgeois Utopias: The Rise and Fall of Suburbia*, New York.

Fontenelle, Bernard Le Bovier, 1972 edition, *Conversations on the Plurality of Worlds*, New York.

Forman, M. 1998, *Nationalism and the International Labor Movement: The Idea of the Nation in Socialist and Anarchist Theory*, University Park, Pa.

Foster, J. 1974, *Class Struggle in the Industrial Revolution*, London.

Foucault, M. 1973 edition, *The Order of Things: The Archaeology of the Human Sciences*, New York.

Foucault, M. 1984, 'Preface' to G. Deleuze and F. Guattari, *Anti-Oedipus: Capitalism and Schizophrenia*, London.

Foucault, M. 1986, 'Of other spaces,' *Diacritics*, 16(1), 22–7.

Foucault, M. 1995 edition, *Discipline and Punish: The Birth of the Prison*, New York, XI–XIV.

Frank, A. 1969, *Capitalism and Underdevelopment in Latin America*, New York.

Frank, A. 1997, 'Quantum bees,' *Discover*, November 1997, 81–7.

Frankel, B. 1987, *The Post-industrial Utopians*, Oxford.

Fraser, N. 1997, *Justice Interruptus*, London.

Friedman, T. 1996, 'Revolt of the Wannabes,' *New York Times*, January 7th, 1996, A19.

Fukuyama, F. 1992, *End of History and the Last Man*, New York.

Geddes, P. 1968, *Cities in Evolution*, New York.

Geras, N. 1983, *Marx and Human Nature: Refutation of a Legend*, London.

Geronimus, A. T., Bound, J., Waidman, T., Hillemeier, M., and Burns, P. 1996, 'Excess mortality among blacks and whites in the United States,' *New England Journal of Medicine*, 335, No. 21, 1552–8.

Gibson, W. 1984, *Neuromancer*, New York.

Glickman, L. 1997, *A Living Wage: American Workers and the Making of Consumer Society*, Ithaca, New York.

Goldberg, D. 1993, *Racist Culture*, Oxford.

Goodman, E. 1996, 'Why not a labor label?' *Baltimore Sun*, July 19th, 1996, 25A.

Gould, S. 1988, *Time's Arrow, Time's Cycle*, New York.

Gramsci, A. 1971 edition, *Selections from the Prison Notebooks*, London.

Gramsci, A. 1978 edition, *Selections from Political Writings, 1921–1926*, London.

Gray, J. 1998, *False Dawn: The Illusions of Global Capitalism*, London.

Greenhouse, S. 1997a, 'Voluntary rules on apparel labor proving elusive,' *New York Times*, February 1st, 1997, 1.

Greenhouse, S. 1997b, 'Accord to combat sweatshop labor faces obstacles,' *New York Times*, April 13th, 1997, 1.

Greider, W. 1997, *One World, Ready or Not*, New York.

Grosz, E. 1994, 'Bodies-cities,' in Colomina, B. (ed.), *Sexuality and Space*, Princeton, 241–53.

Hall, P. 1988, *Cities of Tomorrow*, Oxford.

Hamilton, B., and Kahn, P. 1997, 'Baltimore's Camden Yard Ballpark,' in Noll, R., and Zimbalist, A. (eds), *Sports, Jobs, and Taxes: The Economic Impacts of Sports Teams and Stadiums*, Washington, D.C., 245–81.

Hanke, S. 1996, 'Looks like charity, smells like pork,' *Forbes Magazine*, May, 87.

Hanson, S., and Pratt, G. 1994, *Gender, Work and Space*, London.

Haraway, D. 1991, *Simians, Cyborgs, and Women: The Reinvention of Nature*, London.

Haraway, D. 1995, 'Nature, politics, and possibilities: a debate and discussion with David Harvey and Donna Haraway,' *Society and Space*, 13, 507–27.

Hareven, T. 1982, *Family Time and Industrial Time*, London.

Harvey, D. 1982, *The Limits to Capital*, Oxford.

Harvey, D. 1989, *The Condition of Postmodernity*, Oxford.

Harvey, D. 1992, 'A view from Federal Hill,' in Shopes, L., Fee, E., and Zeidman, L. (eds), *The Baltimore Book: New Views of Local History*, Philadelphia, 226–49.

Harvey, D. 1996, *Justice, Nature and the Geography of Difference*, Oxford.

Hayden, D. 1981, *The Grand Domestic Revolution: A History of Feminist Designs for American Homes, Neighborhoods, and Cities*, Cambridge, Mass.

Hegel, G. W. 1967 edition, *Philosophy of Right*, New York.

Herbert, B. 1997, 'Brutality in Vietnam,' *New York Times*, March 28th, 1997, A29.

Herod, A. 1997, 'Labor as an agent of globalization and as a global agent,' in Cox, K. (ed.), *Spaces of Globalization: Reasserting the Power of the Local*, New York, 167–200.

Herod, A. (ed.), 1998, *Organizing the Landscape: Geographical Perspectives on Labor Unionism*, Minneapolis.

Hetherington, K. 1997, *The Badlands of Modernity: Heterotopia and Social Ordering*, London.

Hirst, G., and Thompson, G. 1996, *Globalization in Question*, Cambridge.

Holmes, S. 1996, 'Children of working poor are up sharply, study says,' *New York Times*, June 4th, 1996, D21.

Huntington, S. 1996, *The Clash of Civilizations and the Remaking of the World Order*, Cambridge, Mass.

International Labour Office, 1996, *World Employment 1996/97: National Policies in a Global Context*, Geneva.

Jacobs, J. 1961, *The Death and Life of Great American Cities*, New York.

Johnson, C. 1974, *Utopian Communism in France: Cabet and the Icarians*, Ithaca, New York.

Karatani, K. 1995, *Architecture as Metaphor: Language, Number, Money*, Cambridge, Mass.

Kasarda, J. 1995, 'Industrial restructuring and the changing location of jobs,' in Farley, R. (ed.), *State of the Union: America in the 1990s*, Vol. 1, New York, 151–76.

Katz, P. 1994, *The New Urbanism: Toward an Architecture of Community*, New York.

Kuhn, T. 1962, *The Structure of Scientific Revolutions*, Chicago.

Kumar, K. 1987, *Utopia and Anti-Utopia in Modern Times*, Oxford.

Kumar, K. 1991, *Utopianism*, Milton Keynes.

Kunstler, J. 1993, *The Geography of Nowhere*, New York.

Kunstler, J. 1996, *Home from Nowhere: Remaking Our Everyday World for the 21st Century*, New York.

Langdon, P. 1994, *A Better Place to Live: Reshaping the American Suburb*, Amherst, Mass.

Lange, J. 1996, Personal interview, Baltimore, Md.

Lefebvre, H. 1976, *The Survival of Capitalism*, New York.

Lefebvre, H. 1991, *The Production of Space*, Oxford.

Lenin, V. I. 1956 edition, *The Development of Capitalism in Russia*, Moscow.

Lenin, V. I. 1970 edition, *Selected Works* (3 vols), Moscow.

Levins, R., and Lewontin, R. 1985, *The Dialectical Biologist*, Cambridge, Mass.

Levitas, R. 1990, *The Concept of Utopia*, London.

Levitas, R. 1993, 'The future of thinking about the future,' in Bird, J., Curtis, B., Putnam, T., Robertson, G., and Tickner, L. (eds), *Mapping Futures*, London, 257–66.

Lewontin, R. 1982, 'Organism and environment,' in Plotkin, H. (ed.), *Learning, Development and Culture*, Chichester, 151–68.

Lowe, D. 1995, *The Body in Late-capitalist USA*, Durham, N. C.

Lukerman, F., and Porter, P. 1976, 'The geography of utopia,' in Lowenthal, D., and Bowden, M. (eds), *Geographies of the Mind: Essays in Historical Geosophy*, New York, 226–49.

Luxemburg, R., and Bukharin, N. 1972 edition, *Imperialism and the Accumulation of Capital*, New York.

Mandeville, B. 1970 edition, *The Fable of the Bees*, Harmondsworth, Middlesex.

Mao Tsetung 1971, *Selected Readings*, Peking.

Marin, L. 1984, *Utopics: Spatial Play*, London.

Marsh, G. P. 1965 edition, *Man and Nature: Or, Physical Geography as Modified by Human Action*, Cambridge, Mass.

Martin, E. 1994, *Flexible Bodies*, Boston.

Marx K. 1964 edition, *The Economic and Philosophic Manuscripts of 1844*, New York.

Marx, K. 1965 edition, *Wages, Price and Profit*, Peking.

Marx, K. 1967 edition, *Capital* (3 vols), New York.

Marx, K. 1968 edition, *Theories of Surplus Value* (3 vols), New York.

Marx, K. 1970 edition, *A Contribution to the Critique of Political Economy*, New York.

Marx, K. 1973 edition, *Grundrisse*, New York.

Marx, K. 1976 edition, *Capital*, Vol. 1, New York.

Marx, K. 1978 edition, *Capital*, Vol. 2, New York.

Marx, K., and Engels, F. 1952 edition, *Manifesto of the Communist Party*, Moscow.

Marx, K., and Engels, F. 1972 edition, *The Marx-Engels Reader* (edited by Robert Tucker), New York.

Marx, K., and Engels, F. 1980 edition, *Collected Works*, Vol. 16, New York.

Marx, K., and Lenin, V. I. 1940 edition, *The Civil War in France: The Paris Commune*, New York.

Meszaros, I. 1995, *Beyond Capital*, New York.

Miyoshi, M. 1997, 'A borderless world,' *Politics-Poetics Documenta X* (the book of the exhibition), Kassel, 182–202.

Moody, K. 1997, *Workers in a Lean World*, London.

More, T. 1901 edition, 'Utopia,' in *Ideal Commonwealths*, London.

Mumford, L. 1961, *The City in History*, New York.

Munn, N. 1985, *The Fame of Gawa*, Cambridge.

Naess, A., and Rothenberg, D. 1989, *Ecology, Community and Lifestyle*, Cambridge.

Nussbaum, M. 1996, *For Love of Country: Debating the Limits of Patriotism*, Boston.

Ohmae, K. 1995, *The End of the Nation State: The Rise of Regional Economies*, London.

Ollman, B. 1993, *Dialectical Investigations*, London.

Ollman, B. (ed.), 1998, *Market Socialism: The Debate among Socialists*, London.

Parfitt, D. 1986, *Reasons and Persons*, Oxford.

Park, R. 1967 edition, *On Social Control and Collective Behavior*, Chicago.

Phillips, A., and Rosas, A. 1995, *Universal Minority Rights*, Abo and London.

Polanyi, K. 1957 edition, *The Great Transformation*, New York.

Pollin, R., and Luce, S. 1998, *The Living Wage: Building a Fair Economy*, New York.

Poovey, M. 1998, *A History of the Modern Fact*, Chicago.

Prigogyne, I., and Stengers, I. 1984, *Order out of Chaos: Man's New Dialogue with Nature*, New York.

Putnam, R. 1993, *Making Democracy Work: Civic Traditions in Modern Italy*, Princeton, N.J.

Readings, B. 1996, *The University in Ruins*, Cambridge, Mass.

Ricoeur, P. 1992, *Oneself as Another*, Chicago.

Robinson, K. 1993–6, *The Mars Trilogy*, New York.

Rodney, W. 1981 edition, *How Europe Underdeveloped Africa*, Washington, D.C.

Ronen, R. 1994, *Possible Worlds in Literary Theory*, Cambridge.

Ross, A. (ed.) 1997, *No Sweat*, London.

Ross, K. 1988, *The Emergence of Social Space: Rimbaud and the Paris Commune*, Minneapolis.

Rothman, D. 1971, *The Discovery of the Asylum*, Boston.

Said, E. 1978, *Orientalism*, New York.

Sandercock, L. 1998, *Towards Cosmopolis*, New York.

Scully, V. 1994, 'The architecture of community,' in Katz, P., *The New Urbanism: Toward an Architecture of Community*, New York.

Schwab, K., and Smadja, C. 1999, 'Globalization needs a human face,' *International Herald Tribune*, January 28th, 1999, 8.

Seabrook, J. 1996, *In the Cities of the South: Scenes from a Developing World*, London.

Sen, A. 1982, *Poverty and Famines*, Oxford.

Sennett, R. 1970, *The Uses of Disorder: Personal Identity and City Life*, New York.

Simmel, G. 1971, 'The metropolis and mental life,' in Levine, D. (ed.), *On Individuality and Social Forms*, Chicago, 324–39.

Smith, A. 1937 edition, *The Wealth of Nations*, The Modern Library, New York.

Smith, H. 1998, 'How the middle class can share in the wealth,' *New York Times*, April 19th, 1998, B18.

Smith, N. 1990, *Uneven Development: Nature, Capital and the Production of Space*, Oxford.

Smith, N. 1992, 'Geography, difference and the politics of scale,' in Doherty, J., Graham, E., and Malek, M. (eds), *Postmodernism and the Social Sciences*, London, 57–79.

Soros, G. 1996, 'The capitalist threat,' *The Atlantic Monthly*, September, 1996, 18–28.

Stafford, B. 1991, *Body Criticism: Imaging the Unseen in Enlightenment Art and Medicine*, Cambridge, Mass.

Strathern, M. 1988, *The Gender of the Gift*, Berkeley.

Swyngedouw, E. 1997, 'Neither global nor local: "glocalization" and the politics of scale,' in Cox, K. (ed.), *Spaces of Globalization: Reasserting the Power of the Local*, New York, 137–66.

Tafuri, M. 1976, *Architecture and Utopia*, Cambridge, Mass.

Thomas, P. 1985, *Karl Marx and the Anarchists*, London.

Thompson, E. P. 1968, *The Making of the English Working Class*, Harmondsworth, Middlesex.

Thornberry, P. 1995, 'The UN Declaration on the rights of person belonging to national, ethnic, religious and linguistic minorities: Background, analysis, observations, and an update,' in Phillips, A., and Rosas, A. (eds), *Universal Minority Rights*, Abo and London, 13–76.

Unger, R. 1987a, *False Necessity: Anti-necessitarian Social Theory in the Service of Radical Democracy*, Cambridge.

Unger, R. 1987b, *Social Theory: Its Situation and Its Task*, Cambridge.

Union of Concerned Scientists, 1996, *World Scientists' Warning to Humanity*, Union of Concerned Scientists, Cambridge, Mass.

United Nations Development Program, 1996, *Human Development Report, 1996*, New York.

Von Frisch, K. 1967, *The Dance Language and Orientation of Bees*, London.

Wallerstein, I. 1974, *The Modern World System*, New York.

Webber, M., and Rigby, D. 1996, *The Golden Age Illusion: Rethinking Postwar Capitalism*, New York.

White, J. B. 1990, *Justice as Translation: An Essay in Cultural and Legal Criticism*, Chicago.

Whitehead, A. N. 1969, *Process and Reality*, New York.

Williams, R. 1989, *Resources of Hope*, London.

Wilson, E. 1998, *Consilience: The Unity of Knowledge*, New York.

Wilson, W. J. 1996, *When Work Disappears: The World of the New Urban Poor*, New York.

World Bank 1995, *World Development Report: Workers in an Integrating World*, New York.

Wright, M. 1996, *Third World Women and the Geography of Skill*, Ph.D. Dissertation, The Johns Hopkins University, Baltimore.

Zeldin, T. 1994, *An Intimate History of Humanity*, New York.

Zola, E. 1891, *Money*, Gloucestershire.

Index